Mechanisms and Machine Science

Volume 41

Series editor

Marco Ceccarelli
LARM: Laboratory of Robotics and Mechatronics
DICeM; University of Cassino and South Latium
Via Di Biasio 43, 03043 Cassino (Fr), Italy
e-mail: ceccarelli@unicas.it

More information about this series at http://www.springer.com/series/8779

Oriol Bohigas · Montserrat Manubens
Lluís Ros

Singularities of Robot Mechanisms

Numerical Computation and Avoidance Path Planning

Springer

Oriol Bohigas
Institut de Robòtica i Informàtica Industrial (UPC-CSIC)
Barcelona, Catalonia
Spain

Lluís Ros
Institut de Robòtica i Informàtica Industrial (UPC-CSIC)
Barcelona, Catalonia
Spain

Montserrat Manubens
Institut de Robòtica i Informàtica Industrial (UPC-CSIC)
Barcelona, Catalonia
Spain

ISSN 2211-0984 ISSN 2211-0992 (electronic)
Mechanisms and Machine Science
ISBN 978-3-319-32920-8 ISBN 978-3-319-32922-2 (eBook)
DOI 10.1007/978-3-319-32922-2

Library of Congress Control Number: 2016942007

Printed on acid-free paper

This Springer imprint is published by Springer Nature
The registered company is Springer International Publishing AG Switzerland

To our families

Preface

Motivation and Purpose

Robotics is flourishing. Innovative robot mechanisms constantly see the light of day, and their use may increase dramatically in the near future. Whether on Earth or in Space, from research labs, to medicine, or industry, we see parallel and walking robots, force-feedback devices, flying manipulators, anthropomorphic hands and arms, humanoids, and other sophisticated machines in action. The capacity to perform complex motions in a precise and reliable way is essential for such devices, but the analysis and planning of such motions is by no means trivial. It requires a deep understanding and treatment of so-called *singular* configurations, the special postures of the mechanism in which kinetostatic behavior degrades to a huge extent.

The purpose of this book is to provide a consistent presentation of all singularity types that can be encountered in a mechanism, together with general robust methods for their interpretation, computation, and avoidance path planning. Obtaining such methods is crucial, because singularities generally pose problems for the normal operation of a robot, and thus they must be taken into account before the actual construction of a prototype. The book also shows how the computation of the singularity set provides detailed information on the local and global motion capabilities of a mechanism: its projections onto the task and joint spaces determine the working regions in such spaces, may inform about the existence of different assembly configurations, and highlight areas where control or dexterity losses arise. These projections also supply a fair view of the feasible movements of the mechanism, but do not reveal all possible singularity-free motions. In order to also tackle this issue, the book develops general path planning methods that avoid problematic singularities, and extends such methods to consider wrench-feasibility constraints in rigid-limbed or cable-driven hexapods.

Although the key role played by singular configurations has been known for years, methods for singularity set computation or avoidance have only been designed for specific mechanisms to date. A distinguishing feature of the book is that it proposes, for the first time, a toolbox of methods applicable to nonredundant mechanisms of large generality, thus facilitating the development of new or more complex robot designs. Emphasis is put on mechanisms with nonredundant actuation because these allow an easier, more symmetric presentation of results, but the techniques, duly extended, could be used in redundant cases as well. In sum, the work seeks to eliminate barriers in design creativity, and to contribute to the general understanding on how the motions of complex multibody systems can be predicted, planned, and controlled in an efficient and reliable way.

Highlights

Overall, the book:

- Makes an effort to provide a clear definition and interpretation of all singularity types, whose descriptions have often been confusing, and very much dispersed in paper publications over the years.
- Provides solutions to two open problems of robot kinematics: the exhaustive computation of the singularity set, and the planning of singularity-free paths, on nonredundant mechanisms of general architecture.
- Shows how the ability to compute the singularity set yields a general method for workspace determination.
- Disseminates powerful branch-and-prune and higher-dimensional continuation methods in the fields of robotics and mechanism science.
- Develops an innovative method for planning wrench-feasible paths for hexapodal parallel kinematic machines.
- Illustrates the performance of the methods in robotic devices of practical interest, including planar, spherical, and spatial manipulators with closed-kinematic chains, and in challenging cable-driven robots.
- Makes intensive use of drawings and pictures to facilitate the reader's understanding of the subject.

The algorithms in the book are distributed within the CUIK Suite, an open-source software package developed by the Kinematics and Robot Design Group at IRI (CSIC-UPC), Barcelona [1]. This package can be downloaded from http://www.iri.upc.edu/cuik, and it can be used to replicate any of the test cases analyzed. The equation files needed to run such test cases can be found in the companion web page of the book http://www.iri.upc.edu/srm, which provides further complementary material, including videos, animated versions of several figures, and an introductory tutorial.

The results of the book were presented by the first author in the Advanced School on "Singular Configurations of Mechanisms and Manipulators," held on September 22–26, 2014 in the International Center for Mechanical Sciences in Udine, Italy [2]. The presentation included a hands-on session in which the attendees could experiment with the algorithms of the book, applying them to simple and illustrative mechanisms. The material of that session is available in http://www.iri.upc.edu/srm, and it can be used as an additional source of information.

Intended Audience

The book is the outcome of several years of collaborative research by the authors, primarily in the context of the Ph.D. work by the first author, supervised by the second and third authors. The style is that of a research monograph, but care has been taken to make the book accessible to a wide audience, from graduate students to professional engineers, mathematicians, or researchers working in robotics, mechanism design, or related fields. The book is also a source of supplementary material for robotics courses at graduate level, especially those devoted to robot kinematics. Altogether, the results may have an impact in a variety of domains, including the design of new parallel robots, haptic devices, 3D printers, mechatronic prostheses, or bio-inspired robots. They may also find application in emerging fields such as programmable surfaces for Earth and Space applications.

Acknowledgments

Several people have contributed to successfully complete this project. First of all, we owe a large debt to Josep M. Porta for his careful and rigorous work on the CUIK Suite over the years. The algorithms in the suite provide solid foundations on which this work is based. Without them it would have been impossible to come up with the solutions proposed. Also, Dr. Dimiter Zlatanov from University of Genoa provided invaluable help on the analysis of mechanism singularities presented in Chaps. 2 and 3, and Dr. Michael E. Henderson, from the IBM T.J. Watson Research Center, introduced the authors to the powerful continuation methods employed in Chaps. 5 and 6. The work of Prof. Illian Bonev, from ÉTS Montréal has likewise been particularly inspiring, and we appreciate his useful feedback on an earlier version of the manuscript. On the hardware side, Alejandro Rajoy and Patrick Grosch did the excellent job of designing and constructing some of the prototypes shown. They worked hard on the parallel 3-RRR robot of Chap. 5, and Patrick also developed the cable-driven robot of Chap. 6. Alex, moreover, produced beautiful illustrations for the book, including the one in this preface.

Our sincere thanks to Prof. Federico Thomas for being always a source of inspiration, support, and guidance, and to Prof. Jorge Angles who, together with Prof. Thomas, encouraged the authors to publish this work as a book. We also thank Prof. Marco Ceccarelli for welcoming the project enthusiastically in the Springer series on Mechanism and Machine Science, and Ms. Anneke Pot from Springer for helping with valuable assistance during the whole editorial process. The help of

Helen Jones in proofreading initial parts of the manuscript is also greatly appreciated.

Our warmest thanks finally go to Gemma, Javier, Núria, and to the rest of our families, for their love, wholehearted help, and encouragement at all moments. We dedicate the book to all of them.

This work has been partially funded by the Spanish Ministry of Economy under the projects DPI2010-18449, DPI2014-57220-C2-2-P, by the CSIC project 201250E026, and by a Juan de la Cierva fellowship supporting the second author.

Barcelona, Catalonia
December 2015

Oriol Bohigas
Montserrat Manubens
Lluís Ros

References

1. Porta, J., Ros, L., Bohigas, O., Manubens, M., Rosales, C., & Jaillet, L. (2014) The CUIK suite: Motion analysis of closed-chain multibody systems. *IEEE Robotics and Automation Magazine*, *21*(3), 105–114.
2. "Advanced School on Singular Configurations of Mechanisms and Manipulators." http://www.cism.it/courses/C1413/. Accessed 26 Dec 2015.

Contents

Notation

Scalars, Vectors, and Tuples

x	A scalar variable, typically a coordinate
t	The time parameter
θ	An angle
v	A linear velocity
ω	An angular velocity
$\mathbf{x}$	A tuple or vector (understood by context). When $\mathbf{x}$ is a vector appearing in an operation, it should be thought of as a column vector
$\mathbf{x}^\mathsf{T}$	The transpose of vector $\mathbf{x}$
$\mathbf{x}(t)$	A time-dependent vector function
$\dot{\mathbf{x}}(t)$	The time derivative of $\mathbf{x}(t)$. Sometimes the dependency on t is omitted
$\mathbf{0}$	A column vector of zeros

Vector Components and Compound Vectors

Whenever the components of a vector need to be made explicit in a displayed formula, they will be written in a column with square brackets, as in

$$\mathbf{v} = \begin{bmatrix} v_1 \\ \vdots \\ v_n \end{bmatrix}.$$

Inline with the text, however, we shall prefer the notation $\mathbf{v} = (v_1, \ldots, v_n)$ to refer to such components. In a similar way, we shall use $\mathbf{u} = (\mathbf{u}_1, \ldots, \mathbf{u}_n)$ to refer to the column vector $\mathbf{u}$ that results from the linear concatenation of the vectors $\mathbf{u}_1, \ldots, \mathbf{u}_n$. Such a notation is meant to be equivalent to the expression $\mathbf{u} = [\mathbf{u}_1^\mathsf{T}, \ldots, \mathbf{u}_n^\mathsf{T}]^\mathsf{T}$, in which the $\mathbf{u}_i$ are thought of as column vectors.

Screws

$\hat{\mathbf{S}}$	A unit screw in axis coordinates, i.e., in (moment, vector) form
$\hat{\mathbf{w}}$	A wrench in ray coordinates, i.e., in (vector, moment) form
$\hat{\mathbf{T}}$	The twist of the end effector, in axis coordinates

Matrices

$\mathbf{X}$	A matrix
$\mathbf{X}^{\mathsf{T}}$	The transpose of $\mathbf{X}$
$\mathbf{X}^{-1}$	The inverse of $\mathbf{X}$
$\mathbf{X}^{i}$	The submatrix of $\mathbf{X}$ obtained by removing its ith row
$\mathbf{0}$	A vector or a matrix of zeros, understood by context
$\mathbf{0}_{m\times n}$	The $m \times n$ matrix of zeros
$\mathbf{I}$	An identity matrix
$\mathbf{I}_{n\times n}$	The $n \times n$ identity matrix
$\mathbf{D}$	A diagonal matrix
$\mathbf{R}$	A 2×2 or 3×3 rotation matrix

Sets

$\mathcal{A}$	A set. In case of being a manifold, it is explicitly noted
$\partial\mathcal{A}$	The boundary of $\mathcal{A}$
$\mathcal{A}\backslash\mathcal{G}$	Set subtraction: the points of $\mathcal{A}$ except those of $\mathcal{G}$
$\mathcal{A}\times\mathcal{G}$	The Cartesian product of $\mathcal{A}$ and $\mathcal{G}$
$\mathcal{C}$	The configuration space, or C-space, of a mechanism
$\mathcal{T}_{\mathbf{q}}\mathcal{C}$	The tangent space of $\mathcal{C}$ at configuration $\mathbf{q}$
$\mathcal{G}$	The set of points $\mathbf{q}$ of $\mathcal{C}$ in which $\mathbf{\Phi}_{\mathbf{q}}(\mathbf{q})$ is rank-deficient
$\mathcal{S}$	The singularity set of a mechanism
$\mathcal{S}_f$	The set of forward singularities of a mechanism
$\mathcal{C}_{sfree}$	The singularity-free C-space, i.e., $\mathcal{C}\backslash\mathcal{S}$
$[\underline{d},\bar{d}]$	The closed real interval of values x such that $\underline{d}\leq x\leq\bar{d}$
$(\underline{d},\bar{d})$	The open real interval of values x such that $\underline{d}<x<\bar{d}$
$\mathcal{B}$	A box, i.e., a Cartesian product of closed real intervals
$B^{\mathcal{W}}$	A box approximation of set $\mathcal{W}$
$\mathcal{P}$	A polytope
$\mathbb{R}^n$	The n-dimensional vector space over the reals

$\mathbb{Z}$ The set of integer numbers
$SO(m)$ The special orthogonal group in dimension m
$SE(m)$ The special Euclidean group in dimension m

Mechanism Symbols

L_j The jth link of a mechanism
$\mathcal{F}_j$ The reference frame attached to the jth link
$\mathcal{F}_1$ The reference frame of L_1, which usually acts as the absolute frame
$\mathbf{r}_j$ The position vector of the origin of $\mathcal{F}_j$ in the absolute frame
$\mathbf{R}_j$ The rotation matrix providing the orientation of $\mathcal{F}_j$ relative to $\mathcal{F}_1$
J_i The ith joint of a mechanism
ω_i The relative velocity at the ith joint, which can be linear or angular
P A point on a link
$\mathbf{p}^{\mathcal{F}_j}$ The position vector of point P in reference frame $\mathcal{F}_j$
$\mathbf{p}$ The position vector of P in the absolute frame, usually $\mathcal{F}_1$
$\mathbf{q}$ The configuration tuple of a mechanism
$\mathbf{v}$ The tuple of input actuated degrees of freedom of a mechanism
$\mathbf{u}$ The tuple of output coordinates defining the mechanism functionality
$\mathcal{Q}$ The manifold of all possible $\mathbf{q}$ values
$\mathcal{V}$ The manifold of all possible $\mathbf{v}$ values
$\mathcal{U}$ The manifold of all possible $\mathbf{u}$ values
n The dimension of the configuration space $\mathcal{C}$
$\mathbf{m}$ The velocity vector of a mechanism $\mathbf{m} = (\boldsymbol{m}_u, \boldsymbol{m}_v, \boldsymbol{m}_p)$
$\mathbf{m}_\mathrm{u}$ The output velocity components (those of the end effector usually)
$\mathbf{m}_\mathrm{v}$ The input velocity components (those of the actuators)
$\mathbf{m}_\mathrm{p}$ The passive velocity components
$\mathbf{L}$ The coefficients matrix of the velocity equation of the mechanism
$\mathbf{f}$ A vector of actuator forces in a mechanism

Maps

φ A scalar-valued map
φ^{-1} The inverse map of φ
$\boldsymbol{\varphi}$ A vector-valued map
$\boldsymbol{\varphi}^{-1}$ The inverse map of $\boldsymbol{\varphi}$
$\boldsymbol{\Phi}(\mathbf{q})$ The differentiable nonlinear map defining the C-space
$\boldsymbol{\Phi}_\mathbf{q}$ The Jacobian matrix of $\boldsymbol{\Phi}(\mathbf{q})$, whose (i, j) entry is $\partial\Phi_i/\partial q_j$
$\boldsymbol{\Phi}_\mathbf{y}$ The Jacobian matrix of $\boldsymbol{\Phi}(\mathbf{q})$ with respect to the components of $\mathbf{y}$, a subvector of $\mathbf{q}$

$\mathbf{\Phi}\|_{\mathcal{A}}$	The map $\mathbf{\Phi}$ with domain restricted to the set $\mathcal{A}$
$\pi_{\mathbf{u}}$	The projection map on the $\mathbf{u}$ space

Other Symbols

$\|\|\mathbf{x}\|\|$	The norm of vector $\mathbf{x}$
$\mathbf{x} \cdot \mathbf{y}$	The dot product of $\mathbf{x}$ and $\mathbf{y}$
$\mathbf{x} \times \mathbf{y}$	The cross product of $\mathbf{x}$ and $\mathbf{y}$
$\ker(\mathbf{X})$	The kernel of the linear transformation represented by matrix $\mathbf{X}$
$\det(\mathbf{X})$	The determinant of the square matrix $\mathbf{X}$

List of Figures

Chapter 1
Introduction

Singularity analysis is a central topic of robot kinematics. It has as a goal the study of certain configurations, termed *singular* or *critical*, where important changes take place in the kinetostatic performance of a mechanism. Rigidity or dexterity losses arise, and there may appear unresolvable or uncontrollable end-effector forces, among other effects (Figs. 1.1 and 1.2). The study of singularities is thus motivated by a desire to avoid these configurations, but it may be helpful to operate close to them sometimes, such as when handling heavy objects, drilling, or fine-positioning, whenever extreme force or motion transformation ratios may be required.

While many commercial mechanisms are designed not to include singularities in their configuration space, this is not always possible, and the appearance of mechanisms with increasingly complex motions demands general singularity-analysis tools to assist the robot designer. This book provides methods to solve two important problems in this context, on nonredundant lower-pair mechanisms of general architecture:

- **Singularity set computation**: For a given mechanism with designated input and output coordinates, compute the entire singularity set, or any of its relevant subsets.
- **Singularity-free path planning**: Given two feasible configurations of a mechanism, find a movement of the mechanism from one configuration to the other that avoids traversing the singularity set.

The characteristics of each problem pose different challenges and thus demand distinct approaches. Because the singularity set typically has the structure of a general algebraic variety, the computation of all singularities requires a method to obtain *all* real solutions of a system of polynomial equations, even when such solutions exhibit several connected components, positive-dimensional subsets, or nonregular points. In contrast, the planning of singularity-free motions demands that we explore the configurations attainable from a given point, moving through a nonlinear variety that, in general, does not admit a global parameterisation. We shall show that branch-and-prune methods constitute an ideal basis from which to deal with the first problem [4–6], whereas the second problem can be approached with recently-developed tools of higher-dimensional continuation [7–9]. As we shall see, moreover, achieving a good formulation is as critical as choosing an appropriate numerical approach. An important part of the book, thus, will also be devoted to deriving suitable systems

O. Bohigas et al., *Singularities of Robot Mechanisms*,
Mechanisms and Machine Science 41, DOI 10.1007/978-3-319-32922-2_1

Fig. 1.1 Loss of rigidity at a singular configuration. The photographs are taken from an experiment carried out at the Vorarlberg University of Applied Sciences, in which a 3-R$\underline{P}$R mechanism is moved to a singular configuration (*top* sequence) so that the operator can feel the shakiness of the configuration (*bottom* picture). The actuator velocities do not determine the platform velocities in this point, and there is a dual loss in the capacity to withstand external forces. The mechanism gets locked, and it has to be taken to a regular configuration manually to resume normal functioning. The full video of the experiment can be seen in [1]

of equations for the two problems confronted. As an introduction, we next review the historical context from which the two problems emerge (Sect. 1.1), establish the assumptions and scope of the proposed solutions (Sect. 1.2), and provide an overall view of the book structure (Sect. 1.3).

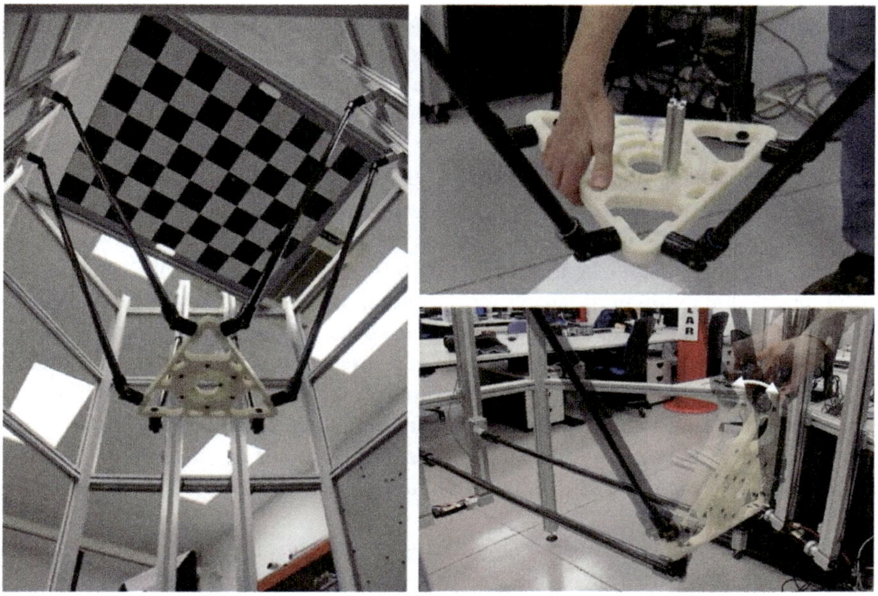

Fig. 1.2 An *hexaglide* mechanism constructed at the Institut de Robòtica i Informàtica Industrial [2]. These robots are typically used as high-speed CNC milling machines [3]. When locking the actuators, the platform is stiff in a nonsingular configuration (*top right*), but it shows uncontrollable motion in a singular one (*bottom right*)

1.1 Historical Context

In a recent editorial [10], McCarthy states that the ability to analyse and design mechanisms and robotic systems of increasing complexity has strongly depended on the ability to solve the associated systems of polynomial equations. It can be expected, he follows, that current advances in computer algebra and numerical continuation methods will spur research in mechanisms and robotics in the coming years. The problems of singularity set computation and singularity-free path planning are clear examples of this situation. Although a proper definition of singularity has existed for a long time, it is only now, after some recent results on polynomial system solving, that such problems have become tractable by extending the current state of the art.

Investigation into singularity analysis started to gain momentum in the early 1980s. The initial efforts were fruitful, especially on serial robot arms, but it soon became clear that a general definition of singularity was needed to facilitate the development of more complex robot designs. Parallel robots of various forms were being developed, and their closed kinematic chains were introducing new phenomena that were difficult to predict and to understand using existing knowledge.

The earliest attempt to classify all singular configurations of a general mechanism can be attributed to Gosselin and Angeles, who used input/output velocity equations to define the well-known Type I and Type II singularities [11]. In essence, these are the configurations in which the velocity of the end-effector does not determine the velocities of the actuators, and *vice versa*. The approach was sound, but neglected the role played by passive joint velocities, and it was later found that further singularity types existed that could not be framed into their formalism [12–16]. This led Zlatanov to define singular configurations in a more general way, as those in which the forward or the inverse instantaneous kinematic problems become unsolvable or indeterminate for some input or output velocities [17]. In appearence, the definition looked similar to that of Gosselin and Angeles, but a key point was the appreciation that these two problems should be understood more generally, as the computation of the *whole* configuration velocity, not just that of the end-effector or the actuators. The new approach, thus, was using the entire configuration space as a central object in the analysis.

Zlatanov's characterisation of singularity is probably the most systematic and general one proposed so far in the literature available, and accommodates, as special cases, the earlier Type I/II singularities, and subtle singularities, such as constraint [18, 19] or architecture singularities [20, 21]. By the end of the 1990s, thus, it was possible to verify whether a given configuration of any mechanism was singular, but it was clear that, for theoretical and practical reasons, it was important to have a means of obtaining the whole singularity set, and to plan mechanism motions avoiding some, or all, of its subsets. As mentioned in [17, Chap. 7], however, significant computational problems remained to be overcome before the emergence of a general method capable of solving the associated systems of equations.

Since the turn of the century, a number of works dealing with singularity set computation or avoidance have appeared, especially in the context of parallel mechanisms. Relevant work in this respect is due to Bonev [22], who studied the singularity sets of several parallel platforms; Mayer St-Onge et al. [23] or Li et al. [24], who provided an analytic form of the six-dimensional singularity set of the general Gough-Stewart platform; Merlet [25], who presented a formal-numerical approach for in-workspace singularity detection; Dasgupta and Mruthyunjaya [26], or Sen et al. [27], who proposed path-deformation or variational techniques for singularity-free path planning; or Jiang and Gosselin [28], who obtained a method to compute the singularity-free orientation workspace of the Gough-Stewart platform. Also with relation to various forms of this platform, Borràs et al. discovered a number of leg rearrangements that leave the singularity set invariant, allowing existing designs with a known singularity locus to evolve to newer, easier-to-construct designs [29–32].

The advances have been significant and we only touch briefly upon them (see [33, 34] for thorough reviews), but the common trend has been to restrict the attention to specific mechanisms, or narrowly-defined classes of mechanisms, and approaches providing a general way to compute or avoid the singularities in arbitrary multibody systems have not been given to date. The purpose of this book is, precisely, to provide a condensed presentation of recent work aimed at filling such gap [35–49].

1.2 Assumptions and Scope

For the purpose of this book, a robotic mechanism will be a multibody system with designated input and output coordinates, i.e. with a specification of which joints are actuated, and which variables describe the intended task-space functionality. We shall focus our attention on *nonredundant* mechanisms, i.e., those in which the number of inputs, and also the outputs, is equal to the number of degrees of freedom to be governed. Duly extended, the results could be applicable to redundant mechanisms as well, but the developments have a simple symmetry in nonredundant cases, which becomes obscured when redundant actuation is possible. Exceptionally, however, our workspace determination methods will be applicable to both types of systems.

Except where noted, the various bodies, or links, of the system will be rigid and interconnected through lower-pair joints, and the connection pattern may be arbitrary, including open and/or closed kinematic chains. Depending on the space where the links operate, such systems are referred to as *planar*, *spherical*, or *spatial* mechanisms. While planar and spherical mechanisms usually have three degrees of freedom (two translations and one rotation, and three rotations, respectively), spatial mechanisms usually have six (three translations and three rotations). Various architectures may be encountered in practice, such as those in the following double page figure, and our results will be applicable to any one of them.

We shall neglect link-link collisions when computing the singularity sets. Although this is a typical assumption in the literature, we note that the consideration of collision constraints would certainly be useful in workspace determination. These constraints, however, add considerable complexity to the problems we consider, and it is more plausible to deal with them a posteriori, using collision checkers like those employed in probabilistic path planning methods. Collision constraints, however, will be tractable by any of the path planners proposed (using penalty terms in the cost functions assumed), and mechanical joint limits will be accommodated by any of the algorithms in the book, by means of suitable equations to take them into account.

Hexapteron robot for rapid prototyping [50].
(Courtesy of Prof. Ilian Bonev, ÉTS Montreal.)

DLR Light Weight Robot arm and Hand [51]
(Courtesy of DLR, Institute of Robotics and Mechatronics, Germany.)

Gough-Stewart platform [52].
(Courtesy of Hydra Power Systems.)

MicARH hexapod for micropositioning [53].
(Courtesy of Prof. Ilian Bonev, ÉTS Montreal.)

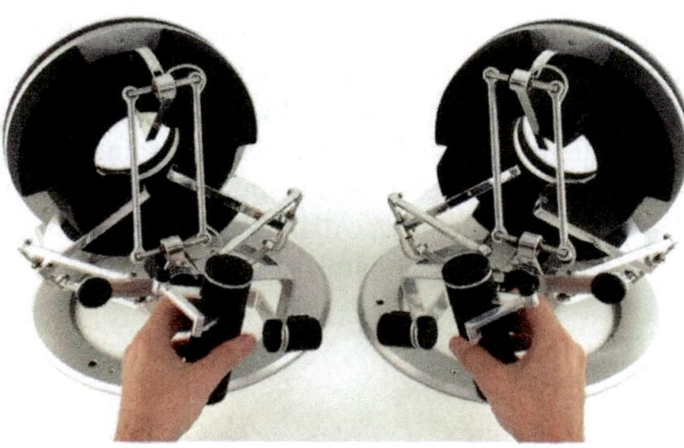

A pair of Omega.7 haptic devices for force feedback [54].
(Courtesy of Force Dimension.)

Exechon X700 parallel kinematic machine [55–57].
(Courtesy of Exechon AB.)

A planar parallel 3-RRR mechanism [58].
(Institut de Robòtica i Inf. Industrial.)

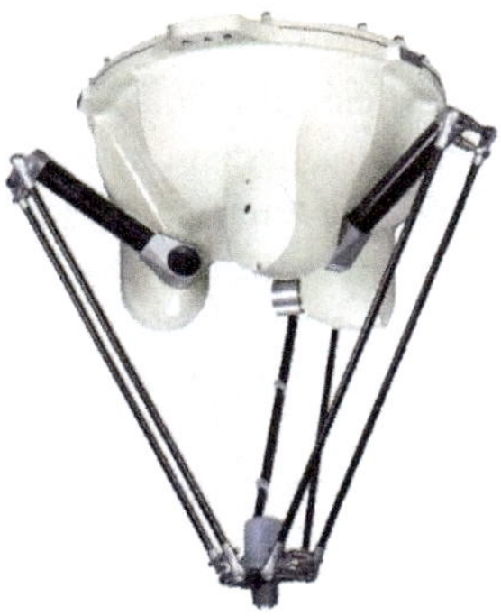

Kawasaki ultra high
speed Delta robot YF03N [59].
(Courtesy of Kawasaki Robotics GmbH.)

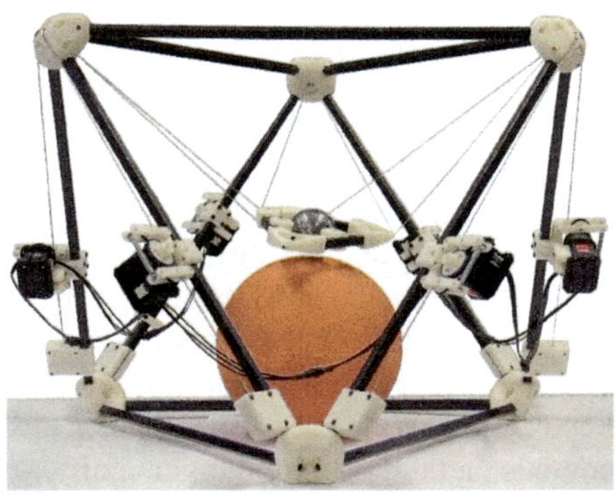

Hexacrane cable-driven platform [40,49].
(Institut de Robòtica i Inf. Industrial.)

Agile Eye, a spherical 3-RRR parallel
robot for the rapid orientation of a camera [60].
(Courtesy of Prof. Clément Gosselin, Laval University.)

1.3 Reader's Guide

The chapter dependencies of the book are outlined in Fig. 1.3. The two diagram branches correspond to methods for singularity set computation and singularity-free path planning, respectively. After reading this chapter and Chap. 2, these two branches can be followed almost independently. The contents of each chapter can be summarised as follows:

This chapter introduces the research, its main objectives and scope, the historical context, and the global structure of the book.

Chapter 2 gives a complete view of the main singularity types and the mathematical conditions that characterise the configurations of each type. Using concepts from differential geometry, the chapter gives a geometric interpretation of these conditions and discusses the kinematic consequences of traversing each singularity subset. The definition of Zlatanov's six lower-level singularity types is then recalled, and all singularities are illustrated on a simple mechanism that admits an analytical approach.

Chapter 3 presents a method to compute any of the singularity sets defined in Chap. 2. The method can isolate any of the sets at the desired precision and is very general, since it can be applied to any nonredundant mechanism with lower-pair joints. To do so, we formulate the systems of equations characterising the singularity sets using quadratic polynomials, and then exploit the special form of such polynomials to define a branch-and-prune scheme able to compute all singularity points. Since the singularity sets are usually defined in a highly-dimensional space, we also provide hints on how to represent these sets in a meaningful and informative way.

Chapter 4 extends the developments in Chaps. 2 and 3 to allow the computation and representation of the various workspaces of a mechanism. The chapter shows how the computation of certain generalised singularities leads to identifying the boundary of a workspace, and the various motion barriers and mobility losses occurring in its interior. A detailed map of the workspace is obtained as a result, indicating all interior and exterior regions and the motion impediments that separate them. Mechanical joint limits are taken into account in the calculations, since they can greatly modify the shape of the workspace. The chapter also compares the proposed method with a previous method of a similar generality.

Chapter 5 provides a singularity-free path planning method. Given two configurations of a mechanism, the method attempts to connect them through a path that does not meet the singularity set at any point. The method can be applied to nonredundant mechanisms of general architecture, and it always provides a path whenever one exists at a given resolution, or determines path nonexistence otherwise. The crucial idea is that the singularity-free region of the configuration space can be placed in correspondence with a smooth manifold, so that by maneuvering through this manifold we can obtain all possible singularity-free movements of the mechanism. To guarantee minimum clearance with respect to the singularity set, the determinant of a Jacobian matrix is obliged to remain larger than a threshold, which is a suitable criterion when only the abstract multibody structure is known.

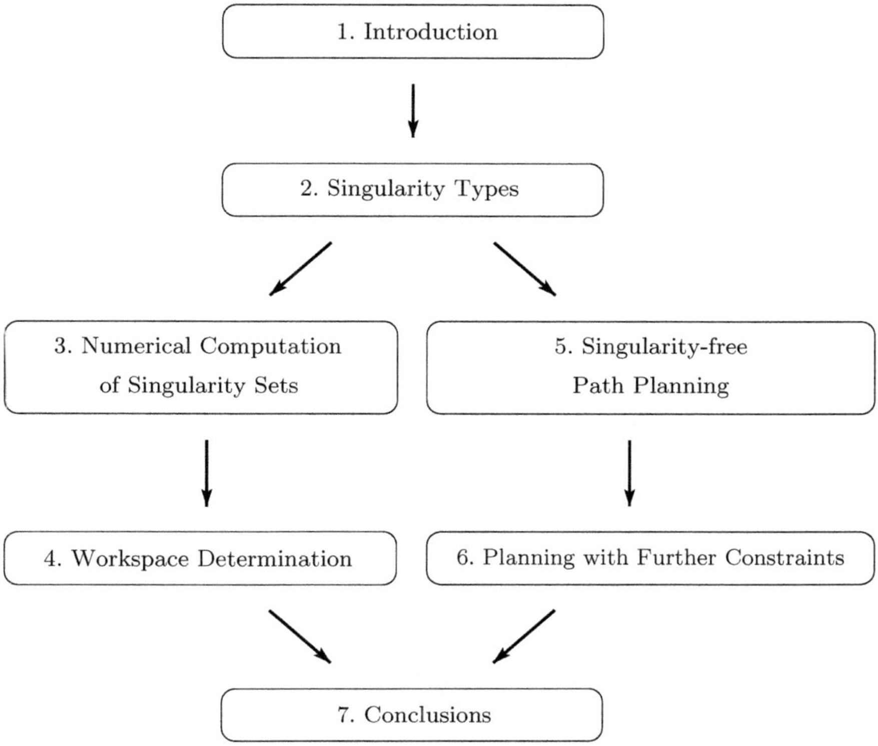

Fig. 1.3 Chapter dependencies

Chapter 6 shows that, when further mechanical properties are known, more meaningful clearance measures can be considered. By way of example, the chapter studies the case of rigid-limbed or cable-driven hexapods, explaining how it is possible to force all configurations to remain wrench-feasible on the computed paths, i.e. able to counteract a given wrench applied to the platform, subject to bounded six-dimensional perturbations. Although the overall study is specific, hints are given on how to extend the developments to also deal with other robot architectures or clearance measures.

Chapter 7, finally, summarises the contributions made in this book and suggests future research directions.

References

1. Companion web page of this book: http://www.iri.upc.edu/srm. Accessed 16 Jun 2016
2. A. Rull, Disseny, implementació i control d'un robot paral·lel de 5 graus de llibertat, Master's thesis, Universitat Politècnica de Catalunya (2011)

3. M. Honegger, A. Codourey, E. Burdet, Adaptive control of the hexaglide, a 6 DOF parallel manipulator, in *Proceedings of the IEEE International Conference on Robotics and Automation, ICRA (Albuquerque, USA)* (1997), pp. 543–548
4. J.M. Porta, L. Ros, T. Creemers, F. Thomas, Box approximations of planar linkage configuration spaces. ASME J. Mech. Design **129**(4), 397–405 (2007)
5. J.M. Porta, L. Ros, F. Thomas, A linear relaxation technique for the position analysis of multi-loop linkages. IEEE Trans. Robot. **25**(2), 225–239 (2009)
6. J. Porta, L. Ros, O. Bohigas, M. Manubens, C. Rosales, L. Jaillet, The CUIK suite: motion analysis of closed-chain multibody systems. IEEE Robot. Autom. Mag. **21**(3), 105–114 (2014)
7. M.E. Henderson, Multiple parameter continuation: computing implicitly defined k-manifolds. Int. J. Bifurcat. Chaos **12**(3), 451–476 (2002)
8. M.E. Henderson, Multiparameter parallel search branch switching. Int. J. Bifurcat. Chaos Appl. Sci. Eng. **15**(3), 967–974 (2005)
9. M.E. Henderson, *Numerical Continuation Methods for Dynamical Systems: Path Following and Boundary Value Problems*, ch. Higher-Dimensional Continuation (Springer, 2007), pp. 77–115
10. J.M. McCarthy, Kinematics, polynomials, and computers–a brief history. ASME J. Mech. Robot. **3**, 7–11 (2011)
11. C.M. Gosselin, J. Angeles, Singularity analysis of closed-loop kinematic chains. IEEE Trans. Robot. Autom. **6**(3), 281–290 (1990)
12. D. Zlatanov, R.G. Fenton, B. Benhabib, Singularity analysis of mechanisms and robots via a motion-space model of the instantaneous kinematics, in *Proceedings of the IEEE International Conference on Robotics and Automation, ICRA (San Diego, USA)* (1994), pp. 980–985
13. D. Zlatanov, R.G. Fenton, B. Benhabib, Singularity analysis of mechanisms and robots via a velocity-equation model of the instantaneous kinematics, in *Proceedings of the IEEE International Conference on Robotics and Automation, ICRA (San Diego, USA)* (1994), pp. 986–991
14. D. Zlatanov, R.G. Fenton, B. Benhabib, Analysis of the instantaneous kinematics and singular configurations of hybrid-chain manipulators, in *Proceedings of the ASME 23rd Biennial Mechanisms Conference (Minneapolis, USA)*, vol. 72 (1994), pp. 467–476
15. F.C. Park, J.W. Kim, Singularity analysis of closed kinematic chains. ASME J. Mech. Design **121**(1), 32–38 (1999)
16. I.A. Bonev, D. Zlatanov, The mystery of the singular SNU translational parallel robot (2001), http://www.parallemic.org. Accessed 26 Dec 2015
17. D. Zlatanov, *Generalized Singularity Analysis of Mechanisms*. Ph.D. thesis, University of Toronto (1998)
18. D. Zlatanov, I.A. Bonev, C. M. Gosselin, Constraint singularities as C-Space singularities, in *Advances in Robot Kinematics: Theory and Applications*, ed. by J. Lenarcic, F. Thomas (Kluwer Academic Publishers, 2002), pp. 183–192
19. D. Zlatanov, I.A. Bonev, C.M. Gosselin, Constraint singularities of parallel mechanisms, in *Proceedings of the IEEE International Conference on Robotics and Automation, ICRA (Washington D.C., USA)*, vol. 1 (2002), pp. 496–502
20. O. Ma, J. Angeles, Architecture singularities of parallel manipulators. Int. J. Robot. Autom. **7**(1), 23–29 (1992)
21. J. Borràs, F. Thomas, C. Torras, Architecture singularities in flagged parallel manipulators, in *Proceedings of the IEEE International Conference on Robotics and Automation, ICRA (Pasadena, USA)* (2008), pp. 3844–3850
22. I.A. Bonev, *Geometric Analysis of Parallel Mechanisms*. Ph.D. thesis, Faculté des Sciences et de Génie, Université de Laval (2002)
23. B.M. St-Onge, C.M. Gosselin, Singularity analysis and representation of the general Gough-Stewart platform. Int. J. Robot. Res. **19**(3), 271–288 (2000)
24. H. Li, C.M. Gosselin, M.J. Richard, B.M. St-Onge, Analytic form of the six-dimensional singularity locus of the general Gough-Stewart platform. ASME J. Mech. Design **128**(1), 279–287 (2006)

25. J.-P. Merlet, A formal-numerical approach for robust in-workspace singularity detection. IEEE Trans. Robot. **23**(3), 393–402 (2007)
26. B. Dasgupta, T.S. Mruthyunjaya, Singularity-free path planning for the Stewart platform manipulator. Mech. Mach. Theory **33**(6), 711–725 (1998)
27. S. Sen, B. Dasgupta, A.K. Mallik, Variational approach for singularity-free path-planning of parallel manipulators. Mech. Mach. Theory **38**(11), 1165–1183 (2003)
28. Q. Jiang, C.M. Gosselin, Determination of the maximal singularity-free orientation workspace for the Gough-Stewart platform. Mech. Mach. Theory **44**(6), 1281–1293 (2009)
29. J. Borràs, F. Thomas, C. Torras, Singularity-invariant families of line-plane 5-SPU platforms. IEEE Trans. Robot. **27**(5), 837–848 (2011)
30. J. Borràs, F. Thomas, C. Torras, On Δ-transforms. IEEE Trans. Robot. **25**(6), 1225–1236 (2009)
31. J. Borràs, *Singularity-Invariant Leg Rearrangements in Stewart-Gough Platforms*. Ph.D. thesis, Universitat Politècnica de Catalunya (2011)
32. J. Borràs, F. Thomas, C. Torras, Singularity-invariant leg rearrangements in doubly-planar Stewart-Gough platforms, in *Robotics: Science and Systems Conference* (2011), pp. 1–8
33. P.S. Donelan, Singularity-theoretic methods in robot kinematics. Robotica **25**(6), 641–659 (2007)
34. P.S. Donelan, *Kinematic Singularities of Robot Manipulators*. InTech (2010)
35. O. Bohigas, L. Ros, M. Manubens, A complete method for workspace boundary determination, in *Advances in Robot Kinematics*, ed. by J. Lenarcic, M. Stanisic (Springer, 2010), pp. 329–338
36. O. Bohigas, L. Ros, M. Manubens, A unified method for computing position and orientation workspaces of general Stewart platforms, in *Proceedings of the ASME International Design Engineering Technical Conferences and Computers and Information in Engineering Conference, IDETC/CIE (Washington D.C., USA)* (2011), pp. 959–968
37. O. Bohigas, L. Ros, M. Manubens, A complete method for workspace boundary determination on general structure manipulators. IEEE Trans. Robot. **28**(5), 993–1006 (2012)
38. O. Bohigas, D. Zlatanov, M. Manubens, L. Ros, On the numerical classification of the singularities of robot manipulators, in *Proceedings of the ASME International Design Engineering Technical Conferences and Computers and Information in Engineering Conference, IDETC/CIE (Chicago, USA)* (2012), pp. 1287–1296
39. J.M. Porta, L. Jaillet, O. Bohigas, Randomized path planning on manifolds based on higher-dimensional continuation. Int. J. Robot. Res. **31**(2), 201–215 (2012)
40. O. Bohigas, M. Manubens, L. Ros, Navigating the wrench-feasible C-space of cable-driven hexapods, in *Cable-Driven Parallel Robots*, ed. by T. Bruckmann, A. Pott (Springer, 2012), pp. 53–68
41. O. Bohigas, M. Manubens, L. Ros, Planning singularity-free force-feasible paths on the Stewart platform, in *Latest Advances in Robot Kinematics*, ed. by J. Lenarcic, M. Husty (Springer, 2012), pp. 245–252
42. O. Bohigas, D. Zlatanov, L. Ros, M. Manubens, J.M. Porta, Numerical computation of manipulator singularities, in *Proceedings of the IEEE International Conference on Robotics and Automation, ICRA (St. Paul, USA)* (2012), pp. 1351–1358
43. O. Bohigas, M.E. Henderson, L. Ros, J.M. Porta, A singularity-free path planner for closed-chain manipulators, in *Proceedings of the IEEE International Conference on Robotics and Automation, ICRA (St. Paul, USA)* (2012), pp. 2128–2134
44. O. Bohigas, M. Manubens, L. Ros, Singularities of non-redundant manipulators: a short account and a method for their computation in the planar case. Mech. Mach. Theory **68**, 1–17 (2013)
45. O. Bohigas, M.E. Henderson, L. Ros, M. Manubens, J.M. Porta, Planning singularity-free paths on closed-chain manipulators. IEEE Trans. Robot. **29**(4), 888–898 (2013)
46. O. Bohigas, M. Manubens, L. Ros, A linear relaxation method for computing workspace slices of the Stewart platform. ASME J. Mech. Robot. **5**, 0011005–1–011005–10 (2013)
47. O. Bohigas, D. Zlatanov, L. Ros, M. Manubens, J.M. Porta, A general method for the numerical computation of manipulator singularity sets. IEEE Trans. Robot. **30**(2), 340–351 (2014)
48. J.M. Porta, L. Ros, O. Bohigas, M. Manubens, C. Rosales, L. Jaillet, An open-source toolbox for motion analysis of closed-chain mechanisms, in *Computational Kinematics*, ed. by F. Thomas, A. Pérez Gracia. Mechanisms and Machine Science, vol. 15 (Springer, 2014), pp. 147–154

49. O. Bohigas, M. Manubens, L. Ros, Planning wrench-feasible motions for cable-driven hexapods. IEEE Trans. Robot. **32**(2), 442–457 (2016)
50. N. Seward, I. Bonev, A new 6-DOF parallel robot with simple kinematic model, in *Proceedings of the IEEE International Conference on Robotics and Automation, ICRA (Hong Kong, China)* (2014), pp. 4061–4066
51. A. Albu-Schaffer, S. Haddadin, C. Ott, A. Stemmer, T. Wimbock, G. Hirzinger, The DLR lightweight robot: design and control concepts for robots in human environments. Ind. Robot: Int. J. **34**(5), 376–385 (2007)
52. Hydra-Power Systems, http://www.hpsx.com/. Accessed 26 Jan 2016
53. J. Coulombe, I.A. Bonev, A new rotary hexapod for micropositioning, in *Proceedings of the IEEE International Conference on Robotics and Automation, ICRA (Karlsruhe, Germany)* (2013), pp. 877–880
54. Force Dimension, http://www.forcedimension.com. Accessed 26 Dec 2015
55. The Exechon Parallel Kinematics Machine, http://www.exechon.com/. Accessed 26 Dec 2015
56. M. Zoppi, D. Zlatanov, R. Molfino, Kinematics analysis of the Exechon tripod, in *Proceedings of the ASME 2010 International Design Engineering Technical Conferences*, Aug 2010, pp. DETC2010–28668 1–8
57. M. Zoppi, D. Zlatanov, R. Molfino, Constraint and singularity analysis of the Exechon tripod, in *Proceedings of the ASME 2010 International Design Engineering Technical Conferences*, Aug 2012, pp. DETC2012–71184 1–10
58. A. Rajoy, Planificación y ejecución de trayectorias libres de singularidades en robots paralelos 3-$\underline{R}$RR, Master's thesis, Universitat Politècnica de Catalunya (2015), http://goo.gl/Hsba1K. Accessed 16 Jun 2016
59. Kawasaki Robotics, http://www.kawasakirobotics.com. Accessed 16 Jun 2016
60. C.M. Gosselin, E. St.-Pierre, Development and experimentation of a fast three-degree-of-freedom camera-orienting device. Int. J. Robot. Res. **16**(5), 619–630 (1997)

Chapter 2
Singularity Types

The literature on singularity analysis is rich, but one often perceives a lack of familiarity with the range of critical phenomena that can arise in a singularity, and with the kind of analytical methods needed to detect such phenomena. To help reversing this trend, and to provide necessary background for the rest of the book, this chapter presents the possible singularity types, their interpretation, and the mathematical conditions characterising each type.

We start from the rigorous definition of singularity given in, which relies [1] on a system of equations defining the feasible velocity vectors of the mechanism (Sect. 2.1). Unlike in other approaches, this vector has enough coordinates to determine the *whole* velocity state, not just the input and output velocities. This is crucial to discover all possible singularity types, some of which would remain unnoticed otherwise. We then provide a novel geometric interpretation of the most common singularities (Sect. 2.2). They can be thought of as C-space *silhouettes* [2], leading to the powerful image in Fig. 2.7, which is quite useful to understand the rest of the book. The image illustrates the main singularity loci, the kinematic consequences of traversing each locus, and how the loci projections yield global views of the potential movements of the mechanism. We also recall Zlatanov's lower-level singularity types, and develop new systems of equations characterising each type (Sect. 2.3). We close the chapter by applying such systems to a simple mechanism with all possible singularities (Sect. 2.4).

2.1 Forward and Inverse Singularities

A mechanism *configuration* can be described by a tuple $\mathbf{q}$ of n_q generalised coordinates which determine the positions and orientations of all links at a given instant of time. There is total freedom in choosing the form and dimension of $\mathbf{q}$, but it must describe one, and only one, configuration. The set of all possible configurations that the mechanism can adopt, then, is called the *configuration space*, or C-space, of the mechanism.

O. Bohigas et al., *Singularities of Robot Mechanisms*,
Mechanisms and Machine Science 41, DOI 10.1007/978-3-319-32922-2_2

Depending on the situation, the values in $\mathbf{q}$ may be independent or not. In the latter case, which is common and usually occurs in closed-chain mechanisms, the tuple $\mathbf{q}$ must satisfy a system of nonlinear equations,

$$\mathbf{\Phi}(\mathbf{q}) = \mathbf{0}, \tag{2.1}$$

which expresses the assembly constraints imposed by the joints [3]. The function $\mathbf{\Phi}(\mathbf{q})$ is a differentiable map

$$\mathbf{\Phi} : \mathcal{Q} \to \mathcal{E},$$

where $\mathcal{Q}$ and $\mathcal{E}$ are n_q- and n_e-dimensional manifolds. Thus, the C-space of the mechanism is the set

$$\mathcal{C} = \{\mathbf{q} \in \mathcal{Q} : \mathbf{\Phi}(\mathbf{q}) = \mathbf{0}\}.$$

We make a distinction between the set $\mathcal{G}$ of points $\mathbf{q} \in \mathcal{C}$ in which the Jacobian

$$\mathbf{\Phi_q}(\mathbf{q}) = \begin{bmatrix} \frac{\partial \Phi_1}{\partial q_1} & \cdots & \frac{\partial \Phi_1}{\partial q_{n_q}} \\ \vdots & & \vdots \\ \frac{\partial \Phi_{n_e}}{\partial q_1} & \cdots & \frac{\partial \Phi_{n_e}}{\partial q_{n_q}} \end{bmatrix}$$

is rank-deficient, and its complement $\mathcal{C} \backslash \mathcal{G}$. By the implicit function theorem [4], $\mathcal{C} \backslash \mathcal{G}$ is a smooth manifold of dimension $n = n_q - n_e$ because we can find a local parameterisation of $\mathcal{C}$ for every $\mathbf{q} \in \mathcal{C} \backslash \mathcal{G}$. The only points in which $\mathcal{C}$ might lose the manifold structure, thus, are those in $\mathcal{G}$, but in most mechanisms $\mathcal{G}$ is of codimension one or higher relative to $\mathcal{C}$.

At a given configuration $\mathbf{q}$, the possible instantaneous motions of the mechanism can be characterised by a system of linear equations,

$$\mathbf{L}\,\mathbf{m} = \mathbf{0}, \tag{2.2}$$

where $\mathbf{L}$ is a matrix that depends on $\mathbf{q}$, and $\mathbf{m}$ is the so-called *velocity vector* of the mechanism. This vector contains enough coordinates to encode the velocities of *all* points of the mechanism, and takes the form

$$\mathbf{m} = \begin{bmatrix} \mathbf{m}_u \\ \mathbf{m}_v \\ \mathbf{m}_p \end{bmatrix},$$

where $\mathbf{m}_u$, $\mathbf{m}_v$, and $\mathbf{m}_p$ encode the output, input, and passive velocities, respectively. Typically, $\mathbf{m}_u$ provides the velocity of a point or the angular velocity, or the twist, of an end-effector body, $\mathbf{m}_v$ provides the actuated joint velocities, and $\mathbf{m}_p$ encompasses the remaining velocity coordinates. Such a system, called the *velocity equation* in [1], can be obtained by differentiation of Eq. (2.1) (provided that $\mathbf{q}$ includes the input and output coordinates, as done in Sect. 2.2), or using twist loop equations (as done in Chap. 3), or by any other means, and therefore it can be used for the practical identification of singularities in general mechanisms. To perform this identification we shall assume that for a given $\mathbf{q}$:

- The *forward* instantaneous kinematic problem (FIKP) consists in finding $\mathbf{m}$, when $\mathbf{m}_v$ is given.
- The *inverse* instantaneous kinematic problem (IIKP) consists in finding $\mathbf{m}$, when $\mathbf{m}_u$ is given.

Note that, contrary to what it is supposed elsewhere [5], in both cases we require to find *all* velocity components of $\mathbf{m}$, not just those referring to the input or output velocities. Following [1], a configuration $\mathbf{q}$ is said to be *nonsingular* when in it both the FIKP and the IIKP have unique solutions for any value of $\mathbf{m}_v$ or $\mathbf{m}_u$; otherwise, it is said to be *singular*. The set $\mathcal{S}$ of all singular configurations is called the *singularity set* of the mechanism.

Two main singularity types can be distinguished in $\mathcal{S}$: *forward* singularities, where the FIKP does not have a unique solution for any value of $\mathbf{m}_v$, and *inverse* singularities, where an analogous indeterminacy occurs in the IIKP relative to $\mathbf{m}_u$.

To characterise the forward and inverse singularity subsets, let us consider the following partitions of the velocity vector

$$\mathbf{m} = \begin{bmatrix} \mathbf{m}_v \\ \mathbf{m}_y \end{bmatrix}, \qquad \mathbf{m} = \begin{bmatrix} \mathbf{m}_u \\ \mathbf{m}_z \end{bmatrix},$$

where $\mathbf{m}_y$ and $\mathbf{m}_z$ encompass the velocities remaining in $\mathbf{m}$ after the removal of $\mathbf{m}_v$ and $\mathbf{m}_u$, respectively. Also let

$$\mathbf{L} = \begin{bmatrix} \mathbf{L}_v \ \mathbf{L}_y \end{bmatrix}, \qquad \mathbf{L} = \begin{bmatrix} \mathbf{L}_u \ \mathbf{L}_z \end{bmatrix},$$

be the corresponding block partitions of $\mathbf{L}$. We can now write Eq. (2.2) in either of the forms

$$\mathbf{L}_v \mathbf{m}_v + \mathbf{L}_y \mathbf{m}_y = \mathbf{0}, \tag{2.3}$$

$$\mathbf{L}_u \mathbf{m}_u + \mathbf{L}_z \mathbf{m}_z = \mathbf{0}. \tag{2.4}$$

Since in Sect. 1.2 we assumed the mechanism to be nonredundant, both the number of coordinates in $\mathbf{m}_v$ and $\mathbf{m}_u$ are equal to the *global mobility* of the kinematic chain, defined as the dimension n of the C-space. In particular, the number of columns of $\mathbf{L}$ exceeds the number of rows by exactly n, and $\mathbf{L}_y$ and $\mathbf{L}_z$ are square matrices [1].

Thus, for configurations $\mathbf{q}$ in which $\mathbf{L}_y$ and $\mathbf{L}_z$ are nonsingular, we can write Eqs. (2.3) and (2.4) in the equivalent forms

$$\mathbf{m}_y = -\mathbf{L}_y^{-1}\, \mathbf{L}_v\, \mathbf{m}_v, \tag{2.5}$$

$$\mathbf{m}_z = -\mathbf{L}_z^{-1}\, \mathbf{L}_u\, \mathbf{m}_u, \tag{2.6}$$

which provide the solution to the FIKP and the IIKP of the mechanism. However, Eqs. (2.5) and (2.6) only hold when $\mathbf{L}_y$ and $\mathbf{L}_z$ are invertible, and only in this case the input and output velocities $\mathbf{m}_v$ and $\mathbf{m}_u$ will determine unique values for the remaining velocities $\mathbf{m}_y$ and $\mathbf{m}_z$. This must be so because, when $\mathbf{L}_y$ is rank deficient at $\mathbf{q}$, Eq. (2.3) yields, for a given value of $\mathbf{m}_v$, either no solution or infinitely-many solutions for $\mathbf{m}_y$, in which case it is not possible to determine the velocity $\mathbf{m}$ of the mechanism by specifying the velocities $\mathbf{m}_v$ of the actuators. When $\mathbf{L}_z$ is rank deficient at $\mathbf{q}$, Eq. (2.4) reveals an analogous relationship between $\mathbf{m}_u$ and $\mathbf{m}$.

Following these observations, note that a configuration $\mathbf{q} \in \mathcal{C}$ is *singular* if, and only if, either $\mathbf{L}_y$ or $\mathbf{L}_z$ is rank deficient at $\mathbf{q}$. Therefore, the singularity set $\mathcal{S}$ can be obtained as the union of the solution sets of the following systems of equations:

$$\left.\begin{array}{r} \mathbf{\Phi}(\mathbf{q}) = \mathbf{0} \\ \mathbf{L}_y\, \boldsymbol{\xi} = \mathbf{0} \\ \|\boldsymbol{\xi}\|^2 = 1 \end{array}\right\}, \tag{2.7}$$

$$\left.\begin{array}{r} \mathbf{\Phi}(\mathbf{q}) = \mathbf{0} \\ \mathbf{L}_z\, \boldsymbol{\xi} = \mathbf{0} \\ \|\boldsymbol{\xi}\|^2 = 1 \end{array}\right\}. \tag{2.8}$$

The first equation in each system constrains $\mathbf{q}$ to be a feasible configuration of the mechanism, and the second and third equations enforce the existence of a nonzero vector $\boldsymbol{\xi}$ in the kernel of the corresponding matrix. Note that $\|\boldsymbol{\xi}\|^2$ can be any consistent norm, for instance $\boldsymbol{\xi}^{\mathsf{T}}\mathbf{D}\boldsymbol{\xi}$ where $\mathbf{D}$ is a diagonal matrix with the proper physical units. There is no need for the norm to be invariant with respect to change of frame or units. In short, the condition $\|\boldsymbol{\xi}\|^2 = 1$ only serves to guarantee that $\boldsymbol{\xi}$ is not $\mathbf{0}$. Clearly, the points $\mathbf{q}$ satisfying the top system are the forward singularities, and those satisfying the bottom system are the inverse singularities. These systems will be very much useful in Chap. 3 to numerically compute the forward and inverse singularity subsets.

When $\mathbf{L}$ is obtained by differentiation of Eq. (2.1), and depending on the formulation adopted, the previous systems of equations may capture formulation singularities [6], e.g. singularities due to a three-parameter parameterisation of $SO(3)$. However, the general formulation proposed in this book avoids this issue by using a redundant representation of $SO(3)$ provided by Eqs. (3.8)–(3.11), and obtaining the velocity equation by means of Screw Theory (Sect. 3.2).

2.2 A Geometric Interpretation of Singularities

Several changes take place when a mechanism traverses a singularity, both at the infinitesimal and finite motion levels. An easy way to understand such changes is provided here using elementary concepts from differential geometry and silhouette analysis [7–9]. To this end, we shall assume that the input and output coordinates of the mechanism appear explicity in $\mathbf{q}$, so that Eq. (2.2) can be directly obtained by taking the time derivative of Eq. (2.1).

2.2.1 *Singularities Yield Shaky Mechanisms*

Consider a point $\mathbf{q} \in \mathcal{C}$, and a C-space trajectory through $\mathbf{q}$, $\mathbf{q}(t)$, where t is a time parameter. From Eq. (2.1) we have

$$\boldsymbol{\Phi}(\mathbf{q}(t)) = \mathbf{0}$$

for all t, because $\mathbf{q}(t)$ lies entirely in the C-space. By differentiating this equation with respect to time we obtain Eq. (2.2) in the form

$$\boldsymbol{\Phi}_{\mathbf{q}}(\mathbf{q})\ \dot{\mathbf{q}}(t) = \mathbf{0}, \tag{2.9}$$

where it becomes apparent that, for a given $\mathbf{q} \in \mathcal{C}$, the feasible velocity vectors of the mechanism are those in the kernel of $\boldsymbol{\Phi}_{\mathbf{q}}(\mathbf{q})$. Assuming the partitions

$$\mathbf{q} = (\mathbf{v}, \mathbf{y}), \qquad \mathbf{q} = (\mathbf{u}, \mathbf{z}),$$

where $\mathbf{v}$ and $\mathbf{u}$ are tuples of n input and n output coordinates, respectively, and $\mathbf{y}$ and $\mathbf{z}$ encompass the remaining coordinates in $\mathbf{q}$, Eq. (2.9) can now be written in either of the forms

$$\boldsymbol{\Phi}_{\mathbf{v}}\ \dot{\mathbf{v}} + \boldsymbol{\Phi}_{\mathbf{y}}\ \dot{\mathbf{y}} = \mathbf{0}, \tag{2.10}$$

$$\boldsymbol{\Phi}_{\mathbf{u}}\ \dot{\mathbf{u}} + \boldsymbol{\Phi}_{\mathbf{z}}\ \dot{\mathbf{z}} = \mathbf{0}, \tag{2.11}$$

where we omit the term dependencies on $\mathbf{q}$ or t for simplicity. Note that in these equations $\boldsymbol{\Phi}_{\mathbf{q}}$, $\boldsymbol{\Phi}_{\mathbf{v}}$, $\boldsymbol{\Phi}_{\mathbf{u}}$, $\boldsymbol{\Phi}_{\mathbf{y}}$, and $\boldsymbol{\Phi}_{\mathbf{z}}$ play the role of $\mathbf{L}$, $\mathbf{L}_{v}$, $\mathbf{L}_{u}$, $\mathbf{L}_{y}$, and $\mathbf{L}_{z}$ in Eqs. (2.2)–(2.4). Thus, a point $\mathbf{q}$ is a forward or an inverse singularity whenever $\boldsymbol{\Phi}_{\mathbf{y}}$ or $\boldsymbol{\Phi}_{\mathbf{z}}$ is rank deficient.

From Eqs. (2.10) and (2.11) we see that if in a singular configuration we lock the input or output coordinates then we obtain an infinitesimally flexible, or shaky, mechanism. Certainly, by setting $\dot{\mathbf{v}} = 0$ and $\dot{\mathbf{u}} = 0$ in Eqs. (2.10) and (2.11), we get homogenous systems of equations with an infinite number of solutions. Such solutions can be interpreted as feasible velocities or infinitesimal movements, and

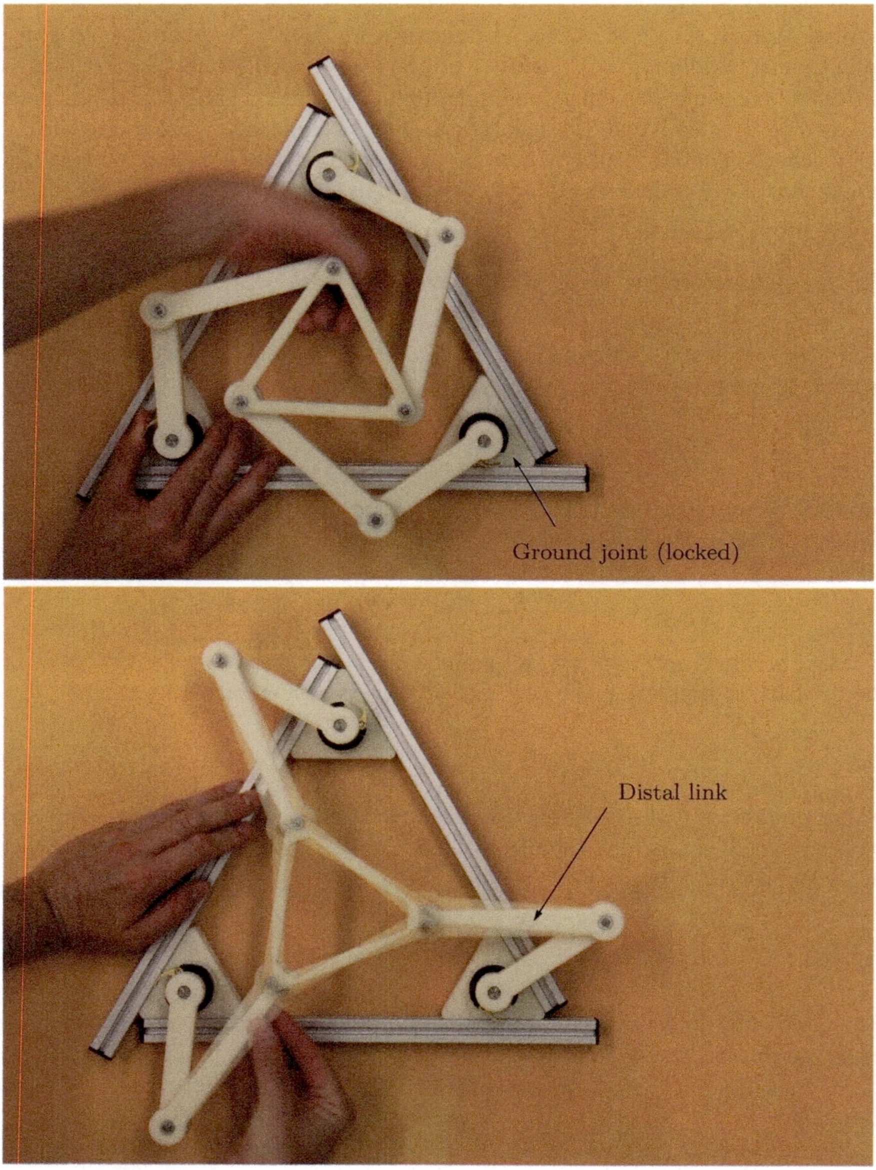

Fig. 2.1 (*Top*) With the ground joints or the platform locked, we feel the mechanism is rigid in a nonsingular configuration. (*Bottom*) In a forward singularity, the distal links are concurrent, and we feel a shakiness after locking the ground joints

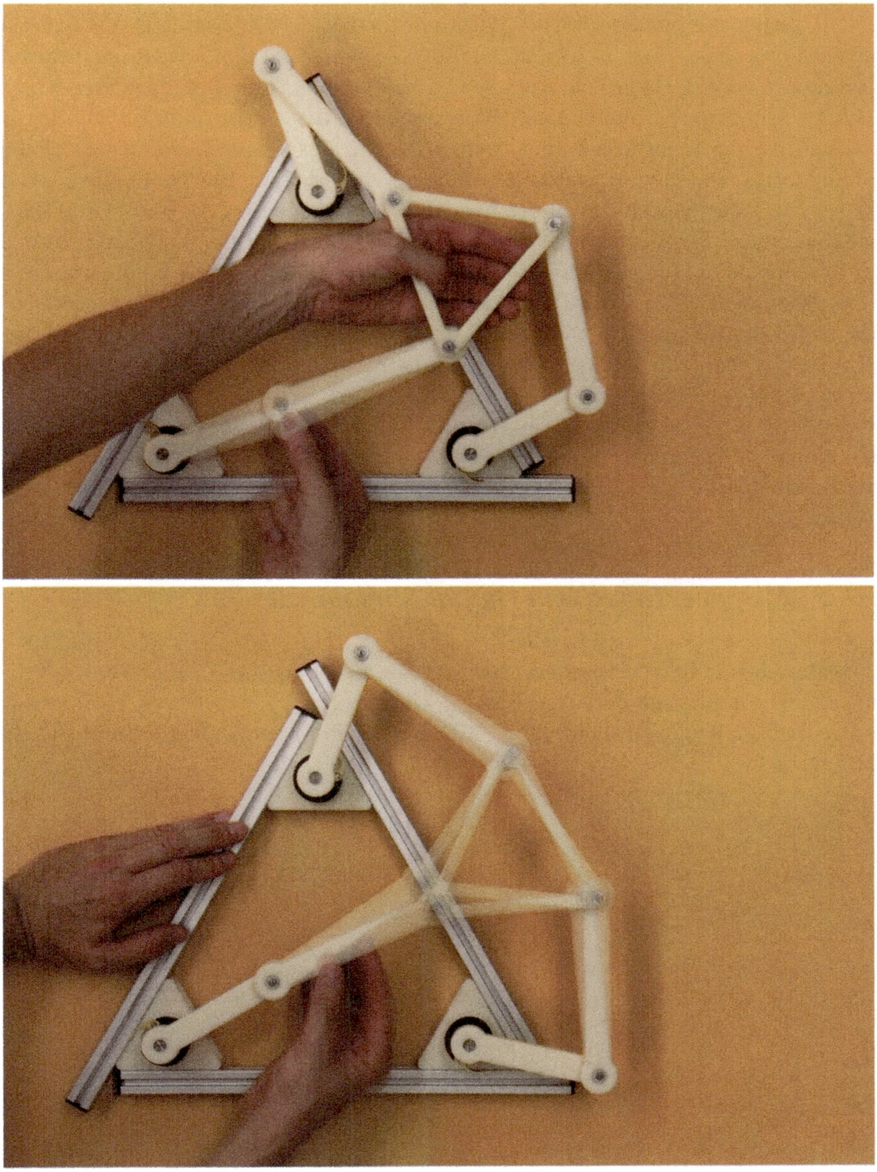

Fig. 2.2 (*Top*) In an inverse singularity, one leg is aligned and, by locking the moving triangle, we feel the leg is shaky. (*Bottom*) a configuration can be both a forward and an inverse singularity at the same time. See [10] for a video of the full sequence

they really arise in an actual mechanism, amplified by mechanical backlash at the joints.

This provides a practical means of testing whether a configuration is singular in a physical prototype. Figures 2.1 and 2.2 illustrate so in a parallel 3-RRR mechanism. This device connects a ground triangle to a moving platform with three legs, where each leg is a three-revolute chain. We assume that the ground joint of each leg is actuated, so the input and output coordinates are, respectively, the angles of the ground joints, and the pose variables of the moving triangle. It is well known that a forward singularity of this mechanism arises when the lines of the three distal links are concurrent, whereas an inverse singularity occurs when the three joints of one leg are aligned [11]. The figures show, respectively, a nonsingular configuration, a forward singularity, an inverse singularity, and a configuration that is both a forward and an inverse singularity. The ground joints are locked in all cases, except in the third one. A video of the full sequence can be seen in the companion web page of this book [10].

In the next two sections we analyse this degenerate behaviour in detail, paying attention to the specific form and location of the tangent space of $\mathcal{C}$, and the kinematic changes that arise as a result.

2.2.2 C-Space, Input, and Output Singularities

Consider all possible curves $\mathbf{q}(t)$ of $\mathcal{C}$ through a point $\mathbf{q} \in \mathcal{C} \backslash \mathcal{G}$, together with the tangent vectors to such curves at $\mathbf{q}$, $\dot{\mathbf{q}}(t)$. The tangent space of $\mathcal{C}$ at $\mathbf{q}$, $\mathcal{T}_{\mathbf{q}}\, \mathcal{C}$, is the space of all such tangent vectors (Fig. 2.3). Since these vectors must fulfill Eq. (2.9), we have

$$\mathcal{T}_{\mathbf{q}}\, \mathcal{C} = \ker(\mathbf{\Phi_q}),$$

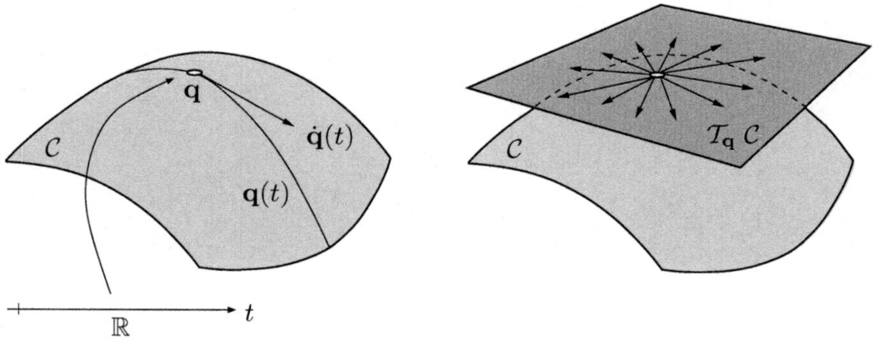

Fig. 2.3 The tangent space at a point $\mathbf{q} \in \mathcal{C} \backslash \mathcal{G}$, $\mathcal{T}_{\mathbf{q}}\, \mathcal{C}$, is formed by the tangent vectors to all possible curves $\mathbf{q}(t)$ through $\mathbf{q}$

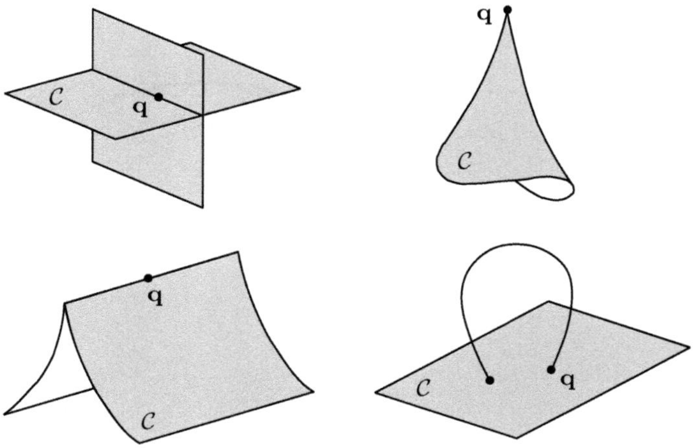

Fig. 2.4 Points $\mathbf{q} \in \mathcal{C}$ where $\mathbf{\Phi_q}$ is rank deficient typically correspond to bifurcations, cusps, ridges, or dimension changes of $\mathcal{C}$

and $\mathcal{T}_\mathbf{q}\, \mathcal{C}$ is a vector space of dimension $n = n_q - n_e$. When $\mathbf{\Phi_q}$ is full rank, thus, the feasible velocity vectors lie in a tangent space of the same dimension as that of $\mathcal{C}$, which is the same as the number of inputs or outputs in a nonredundant mechanism.

If $\mathbf{\Phi_q}$ is rank deficient at some $\mathbf{q} \in \mathcal{C}$, $\mathcal{C}$ may lose the manifold structure. This occurs for example in bifurcations, cusps, ridges, or dimension changes of $\mathcal{C}$ (Fig. 2.4). The tangent space is ill-defined at such points, because some vectors in $\ker(\mathbf{\Phi_q})$ do not correspond to the derivative of any parametric curve of $\mathcal{C}$. However, all vectors of $\ker(\mathbf{\Phi_q})$ still correspond to feasible infinitesimal movements of the mechanism, and can certainly be observed in a real prototype. They form a space of dimension larger than n, because $\mathbf{\Phi_q}$ is not full rank.

In general, one can say that the feasible velocity vectors (or infinitesimal movements) of the mechanism are those in the Zariski tangent space of $\mathcal{C}$ at $\mathbf{q}$, which is defined as $\ker(\mathbf{\Phi_q})$ [1]. The dimension of the Zariski tangent space at some $\mathbf{q} \in \mathcal{C}$ is called the *instantaneous mobility* of the mechanism, and provides the number of independent infinitesimal movements at the given configuration.

Depending on the rank of $\mathbf{\Phi_q}$, $\mathbf{\Phi_y}$, or $\mathbf{\Phi_z}$, one can distinguish three new types of singularities:

- *C-space singularities*: the points $\mathbf{q} \in \mathcal{G} \subset \mathcal{C}$, in which the whole Jacobian $\mathbf{\Phi_q}$ is rank deficient, so that both the FIKP and IIKP become unsolvable or indeterminate independently of the choice of input and output coordinates.
- *Input singularities*: the points $\mathbf{q} \in \mathcal{C} \backslash \mathcal{G}$ where $\mathbf{\Phi_y}$ is rank deficient, so that the FIKP becomes unsolvable or indeterminate in them.
- *Output singularities*: the points $\mathbf{q} \in \mathcal{C} \backslash \mathcal{G}$ where $\mathbf{\Phi_z}$ is rank deficient, so that the IIKP becomes unsolvable or indeterminate.

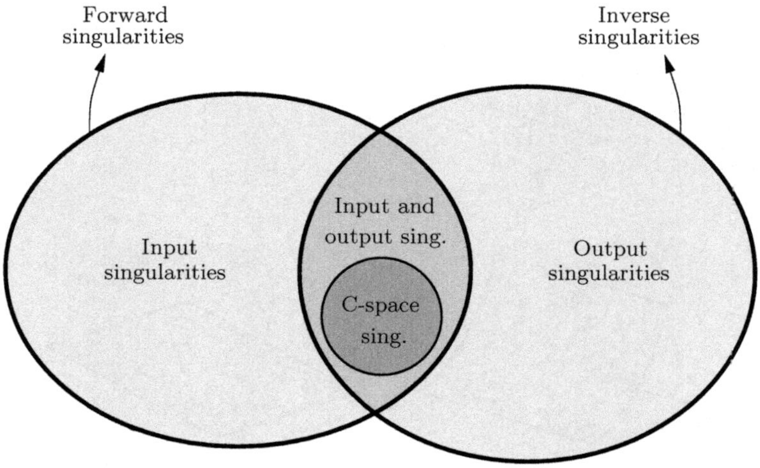

Fig. 2.5 Inclusion relationships of C-space, input, and output singularites, relative to forward and inverse singularities

C-space singularities correspond to the points $\mathbf{q} \in \mathcal{G}$, where $\mathcal{C}$ may lose its manifold structure. Since $\mathbf{\Phi_q}$ is rank deficient in such points, the tangent space of $\mathcal{C}$ is ill-defined and we have an increase of the instantaneous mobility of the mechanism. In input and output singularities, on the contrary, $\mathbf{\Phi_q}$ is full rank, and $\mathcal{C}$ has an n-dimensional tangent space, but this space adopts a very special position. We shall see so in the following section, explaining how this leads to control or dexterity losses of the mechanism.

Note that when $\mathbf{\Phi_q}$ is rank deficient, $\mathbf{\Phi_y}$ and $\mathbf{\Phi_z}$ are rank deficient too, meaning that the forward singularities contain both the input and C-space singularities, whereas the inverse singularities contain both the output and C-space singularities. However, a configuration can be both an input and an output singularity without being a C-space singularity (Fig. 2.5).

2.2.3 Singularities as Silhouette Points

Input and output singularities can be interpreted as "silhouette points" of the C-space, which allows us to understand the kinematic consequences of traversing these points in a visual form. To realise so, we need to recall some background on critical points of differentiable maps [7–9].

Let $\mathbf{\Gamma}(\mathbf{q}) : \mathcal{Q} \rightarrow \mathcal{N}$ be an arbitrary differentiable map between two manifolds $\mathcal{Q}$ and $\mathcal{N}$, of dimensions n_q and n_n respectively. Then, the differential of $\mathbf{\Gamma}$ at $\mathbf{q}$ is the map between tangent spaces

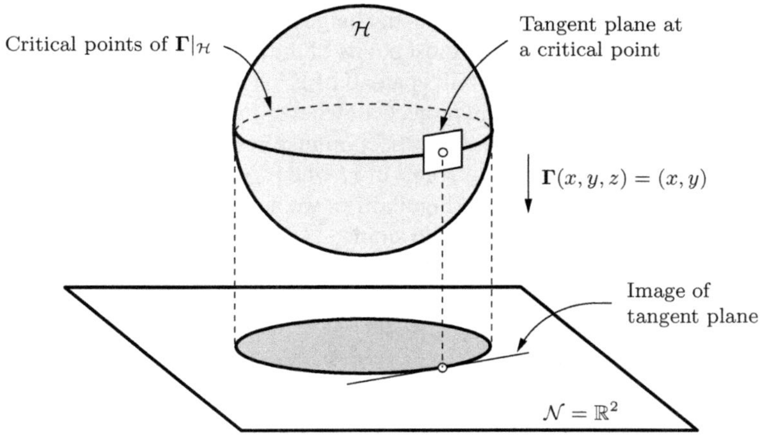

Fig. 2.6 The critical points of the projection of a sphere on $\mathbb{R}^2$ (adapted from [9])

$$\mathbf{\Gamma}_{\mathbf{q}} : \mathcal{T}_{\mathbf{q}}\mathcal{Q} \longrightarrow \mathcal{T}_{\mathbf{\Gamma}(\mathbf{q})}\mathcal{N}$$

induced by the Jacobian matrix $\mathbf{\Gamma}_{\mathbf{q}} = [\frac{\partial \Gamma_i}{\partial q_j}(\mathbf{q})]$. A point $\mathbf{q} \in \mathcal{Q}$ is said to be a *critical point* of $\mathbf{\Gamma}$ if the differential $\mathbf{\Gamma}_{\mathbf{q}}$ is not surjective at $\mathbf{q}$, i.e. if its Jacobian matrix has rank less than n_n at $\mathbf{q}$. Points $\mathbf{q} \in \mathcal{Q}$ that are not critical are called *regular* points. One can also speak of the critical points of $\mathbf{\Gamma}|_{\mathcal{H}}$, the map $\mathbf{\Gamma}$ with domain restricted to a submanifold $\mathcal{H} \subseteq \mathcal{Q}$. These are the points $\mathbf{q} \in \mathcal{H}$ for which $\mathbf{\Gamma}_{\mathbf{q}}(T_{\mathbf{q}}\mathcal{H})$ does not fully span $\mathcal{T}_{\mathbf{\Gamma}(\mathbf{q})}\mathcal{N}$.

The example in Fig. 2.6 serves to illustrate these notions. Suppose that $\mathcal{Q} = \mathbb{R}^3$, $\mathcal{N} = \mathbb{R}^2$, and $\mathbf{q} = (x, y, z)$, and that $\mathbf{\Gamma}$ is the projection map from $\mathbb{R}^3$ to $\mathbb{R}^2$, defined as

$$\mathbf{\Gamma}(x, y, z) = (x, y).$$

We thus have $n_q = 3$ and $n_n = 2$. Also assume that $\mathcal{H}$ is the sphere of $\mathbb{R}^3$ defined implicitly by

$$x^2 + y^2 + z^2 = 1.$$

Note that

$$\mathbf{\Gamma}_{\mathbf{q}} = \begin{bmatrix} 1 & 0 & 0 \\ 0 & 1 & 0 \end{bmatrix},$$

which projects vectors from $\mathbb{R}^3$ to $\mathbb{R}^2$. In this case, the map $\mathbf{\Gamma}_{\mathbf{q}} : \mathbb{R}^3 \rightarrow \mathbb{R}^2$ does not have critical points, because $\text{rank}(\mathbf{\Gamma}_{\mathbf{q}}) = 2 = \dim(\mathbb{R}^2)$ for all $(x, y, z) \in \mathbb{R}^3$. However, if we consider the map $\mathbf{\Gamma}$ with domain restricted to $\mathcal{H}$, $\mathbf{\Gamma}|_{\mathcal{H}}$, critical points do certainly arise. In that case, $\mathbf{\Gamma_q}$ projects the tangent plane of the sphere down to the (x, y) plane. Relative to such map, most points of the sphere are regular, because the projection of their tangent plane will span all of $\mathbb{R}^2$. The equator points are critical though, because the tangent plane projects down to a line of $\mathbb{R}^2$ in them.

In general, the critical points of a projection map restricted to some manifold $\mathcal{H}$ can be interpreted as the silhouette points of $\mathcal{H}$ when "observed from" the projection space [9]. To relate such points to singularities, we need the following result.

Suppose that, as in the previous example, $\mathcal{H}$ is a smooth manifold implicitly defined by a system of n_e equations $\mathbf{\Phi}(\mathbf{q}) = \mathbf{0}$. Then, a point $\mathbf{q} \in \mathcal{H}$ is a critical point of $\mathbf{\Gamma}|_{\mathcal{H}}$ if, and only if, the matrix

$$(\mathbf{\Phi}, \mathbf{\Gamma})_{\mathbf{q}} = \begin{bmatrix} \mathbf{\Phi_q} \\ \mathbf{\Gamma_q} \end{bmatrix}$$

has rank less than $n_e + n_n$ at $\mathbf{q}$ [9, Lemma 4.4.1].

In the case of Fig. 2.6, for example, $\mathbf{\Phi}(\mathbf{q}) = x^2 + y^2 + z^2 - 1,\; n_e = 1$, $\mathbf{\Gamma}(x, y, z) = (x, y)$, and we have

$$\begin{bmatrix} \mathbf{\Phi_q} \\ \mathbf{\Gamma_q} \end{bmatrix} = \begin{bmatrix} 2x & 2y & 2z \\ 1 & 0 & 0 \\ 0 & 1 & 0 \end{bmatrix}.$$

Therefore, the critical points of the projection of the sphere onto the (x, y) plane are those in which the rank of this matrix is less than $n_e + n_n = 3$. This occurs whenever $z = 0$, i.e., for the equator points, as expected.

Now suppose that $\mathcal{H}$ is our C-space $\mathcal{C}$ defined by Eq. (2.1), and let $\mathcal{U}$ and $\mathcal{V}$ respectively denote the spaces of the $\mathbf{u}$ and $\mathbf{v}$ coordinates defined in Sect. 2.2.1. Also assume that

$$\pi_{\mathbf{u}}(\mathbf{q}) : \mathcal{Q} \longrightarrow \mathcal{U}$$

denotes the projection map from $\mathbf{q} = (\mathbf{u}, \mathbf{z})$ onto the $\mathbf{u}$ variables, i.e.,

$$\pi_{\mathbf{u}}(\mathbf{u}, \mathbf{z}) = \mathbf{u}.$$

Then, by the previous result, the critical points of $\pi_{\mathbf{u}}|_{\mathcal{C} \setminus \mathcal{G}}$ are the points $\mathbf{q}$ for which

$$(\mathbf{\Phi}, \pi_{\mathbf{u}})_{\mathbf{q}} = \left[\begin{array}{c:c} \mathbf{\Phi_u} & \mathbf{\Phi_z} \\ \hdashline \mathbf{I}_{n \times n} & \mathbf{0} \end{array}\right]$$

is rank deficient, where $\mathbf{I}_{n\times n}$ denotes the $n \times n$ identity matrix. Observe that $(\mathbf{\Phi}, \pi_{\mathbf{u}})_{\mathbf{q}}$ is rank deficient whenever $\mathbf{\Phi_z}$ is rank deficient, which proves that the output singularities are the mentioned critical points. The same reasoning can be applied to conclude

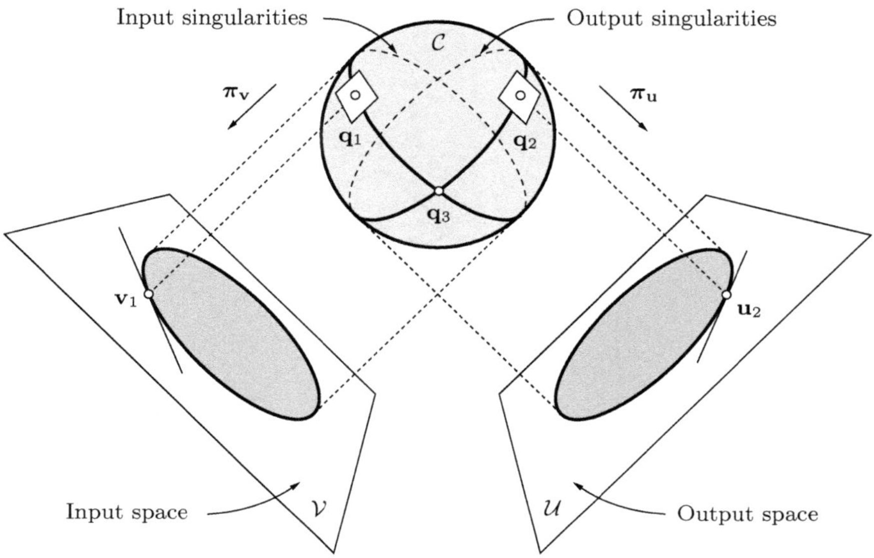

Fig. 2.7 Interpretation of input and output singularities as silhouette points when $\mathcal{Q} = \mathbb{R}^3$, $\mathcal{C}$ is a sphere, and $\mathcal{V}$ and $\mathcal{U}$ are two coordinate planes of $\mathbb{R}^3$. In the figure, $\mathbf{q}_1$ and $\mathbf{q}_2$ correspond to an input and an output singularity, respectively, and $\mathbf{q}_3$ is both an input and an output singularity. A feasible velocity vector at $\mathbf{v}_1$ or $\mathbf{u}_2$ does not locally determine a unique velocity vector in $\mathcal{C}$

that the input singularities correspond to the critical points of the projection map $\pi_{\mathbf{v}}|_{\mathcal{C}\setminus\mathcal{G}}$, where

$$\pi_{\mathbf{v}}(\mathbf{v}, \mathbf{y}) = \mathbf{v}.$$

Singularities, therefore, can be interpreted as silhouettes of the C-space when "seen from" the $\mathcal{U}$ and $\mathcal{V}$ spaces. This is easy to represent when $\mathcal{Q} = \mathbb{R}^{n_q}$, in which case $\mathcal{C}$ can be regarded as a subset of $\mathbb{R}^{n_q}$ (Fig. 2.7). Input and output singularities correspond to points $\mathbf{q}$ where the tangent space of $\mathcal{C}$ projects to $\mathcal{V} = \mathbb{R}^n$ or $\mathcal{U} = \mathbb{R}^n$ as a linear space of dimension lower than n.

Note that input singularities yield uncontrollable motions because a feasible vector $\dot{\mathbf{v}}$ does not determine a unique vector $\dot{\mathbf{q}}$ in them. Output singularities, on the contrary, correspond to dexterity losses of the end-effector. Independently of the value of $\dot{\mathbf{q}}$, $\dot{\mathbf{u}}$ is always restricted to a linear subspace of smaller dimension.

It becomes apparent that input and output singularities determine the boundaries of the input and output workspaces, i.e., the sets of possible values of $\mathbf{v}$ and $\mathbf{u}$ for any $\mathbf{q} \in \mathcal{C}$. We shall develop this point in detail in Chap. 4 using a generalised notion of singularity. Using that notion, we shall provide numerical methods to compute the workspace of any relevant set of coordinates in the mechanism, including $\mathbf{u}$ and $\mathbf{v}$ in particular.

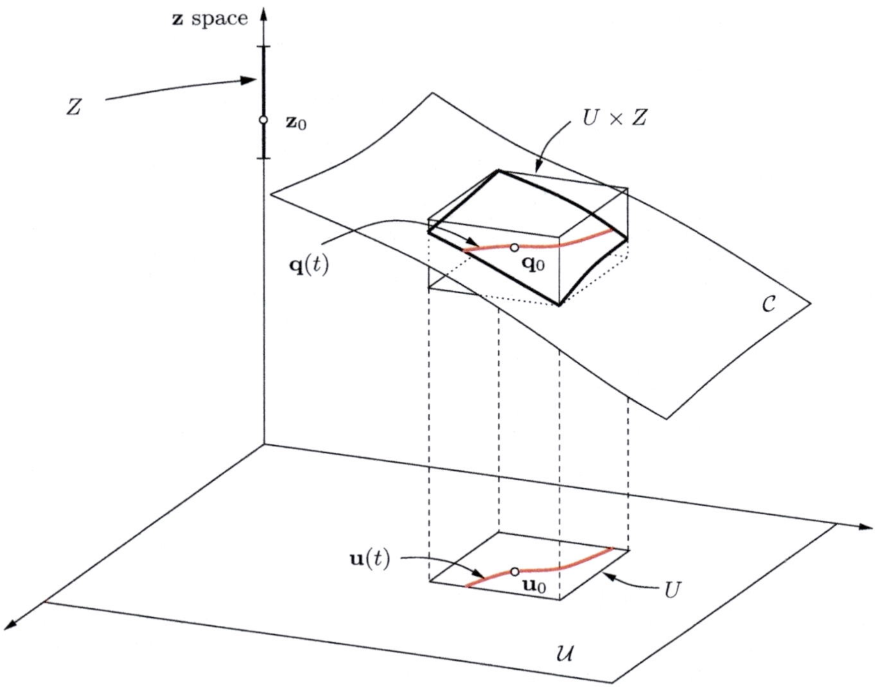

Fig. 2.8 If the configuration $\mathbf{q}_0 = (\mathbf{u}_0, \mathbf{z}_0)$ is not an inverse singularity, we can find a function $\mathbf{z} = \mathbf{F}(\mathbf{u})$, from an open set U around $\mathbf{u}_0$ to the set $Z = F(U)$, such that $\mathbf{q}(\mathbf{u}) = (\mathbf{u}, \mathbf{F}(\mathbf{u}))$ is a valid parameterisation of $\mathcal{C}$ inside $U \times Z$

2.2.4 Indeterminacies in the Mechanism Trajectory

The implicit function theorem provides further insight as to the advantages of avoiding each singularity type [4]. Observe first that, as a consequence of the theorem, if $\boldsymbol{\Phi}_\mathbf{z}$ is full rank at $\mathbf{q}_0 = (\mathbf{u}_0, \mathbf{z}_0) \in \mathcal{C}$, it is possible to find a function $\mathbf{z} = \mathbf{F}(\mathbf{u})$ relating $\mathbf{z}$ with the remaining $\mathbf{u}$ variables, satisfying $\boldsymbol{\Phi}(\mathbf{u}, \mathbf{F}(\mathbf{u})) = \mathbf{0}$ (Fig. 2.8). Thus, the $\mathbf{u}$ variables can be used as a local parameterisation of $\mathcal{C}$ around $\mathbf{q}_0$, implying that arbitrary values in a neighbourhood of $\mathbf{u}_0$ have a corresponding $\mathbf{z}$ satisfying $\boldsymbol{\Phi}(\mathbf{u}, \mathbf{z}) = \mathbf{0}$. This means that a smooth output trajectory $\mathbf{u}(t) \subset \mathcal{U}$ through a point $\mathbf{u}_0$ will locally correspond to a unique smooth trajectory $\mathbf{q}(t)$ in $\mathcal{C}$ through the corresponding $\mathbf{q}_0$, or, in other words, a tracking of the output will be sufficient to predict the overall motion of the mechanism. In a similar way, whenever $\boldsymbol{\Phi}_\mathbf{y}$ is full rank at $\mathbf{q}_0 = (\mathbf{y}_0, \mathbf{v}_0)$, a smooth input trajectory $\mathbf{v}(t)$ through $\mathbf{v}_0$ will locally determine a unique smooth trajectory $\mathbf{q}(t)$ in $\mathcal{C}$, so that the overall movement of the mechanism will be controllable through the inputs.

This one-to-one correspondence between the input or output trajectories, on the one hand, and the mechanism trajectory, on the other hand, is not guaranteed at a

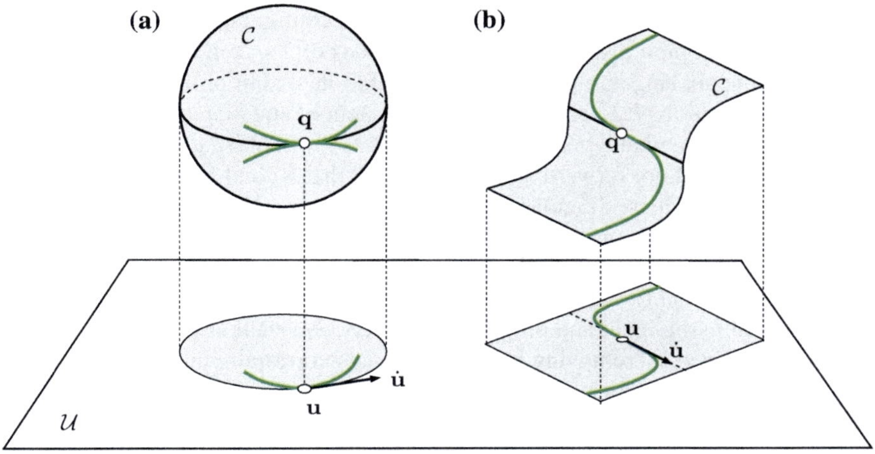

Fig. 2.9 A trajectory in $\mathcal{U}$ through an output singularity $\mathbf{u}$ may determine several trajectories in $\mathcal{C}$ (**a**) or even a single smooth one (**b**), but in all cases the velocity vector $\dot{\mathbf{u}}$ will correspond to infinitely-many velocities $\dot{\mathbf{q}}$, because of the special position of $\mathcal{T}_{\mathbf{q}}\mathcal{C}$

singular configuration, as one can already see in the example of Fig. 2.9a. However, since the implicit function theorem only gives sufficient conditions to guarantee the uniqueness and smoothness of $\mathbf{q}(t)$, points with $\mathbf{\Phi_y}$ or $\mathbf{\Phi_z}$ singular can in principle exist where $\mathbf{v}(t)$ or $\mathbf{u}(t)$ determine a unique smooth trajectory $\mathbf{q}(t)$. An example of this situation is given in Fig. 2.9b. Such points are not exempt of problems though, since $\dot{\mathbf{v}}$ or $\dot{\mathbf{u}}$ will still not determine $\dot{\mathbf{q}}$ in them.

2.3 Lower-Level Singularity Types

We have seen that singular configurations are those in which the forward or the inverse instantaneous kinematic problems of the mechanism become unsolvable or indeterminate. Depending on the kinematic cause of indeterminacy, however, six substantially different types of singularities were recognised by Zlatanov [1]. These are redundant input (RI), redundant output (RO), impossible input (II), impossible output (IO), redundant passive motion (RPM), and increased instantaneous mobility (IIM) singularities.

When the mechanism is in a singularity of RO or IO type, the output velocity is indeterminate or restricted. In contrast, in RI or II type singularities it is the input velocity the one that is indeterminate or restricted. In an RPM type singularity, the passive velocity of the mechanism is indeterminate, which may create problems such as interference with other links and obstacles. In an IIM type singularity, finally, the instantaneous motion of the mechanism is indeterminate, irrespective of which n input parameters are being used to control the movement.

It is therefore desirable to know whether a given configuration belongs to one of the types, or to compute all possible configurations of a specific type. To this end, for each of the six singularity types we first recall its definition from [1] based on the velocity equation (2.2), and then derive a system of equations that characterises the configurations of that type. From the definitions it will become clear that RO and II type singularities are forward singularities, and that RI and IO type singularities are inverse singularities. Singularities of RPM type must lie in the intersection of forward and inverse singularities, and IIM type singularities are exactly the C-space singularities (Fig. 2.5).

Recall that $\mathbf{L}_y$ and $\mathbf{L}_z$ are the submatrices of $\mathbf{L}$ obtained by removing the columns corresponding to the input and output, respectively. We shall also let $\mathbf{L}_p$ denote the submatrix obtained by removing both the columns corresponding to the input and output velocities. In other words, $\mathbf{L} = \left[\mathbf{L}_u \; \mathbf{L}_v \; \mathbf{L}_p\right]$. In the following, we shall use these definitions, and the 3-slider and the 4-bar mechanisms of Fig. 2.10, to illustrate the different singularity types encountered. Each of these mechanisms has one degree of freedom and, unless otherwise stated, the input and output velocities are those of points A and B, v_A and v_B, for the 3-slider mechanism, and the angular velocities of links AB and DC, ω_A and ω_D, for the 4-bar mechanism.

Redundant Input

A configuration is a singularity of RI type if there exists an input velocity vector $\mathbf{m}_v \neq \mathbf{0}$ and a vector $\mathbf{m}_p$ that satisfy the velocity equation (2.2) for $\mathbf{m}_u = \mathbf{0}$, i.e. such that

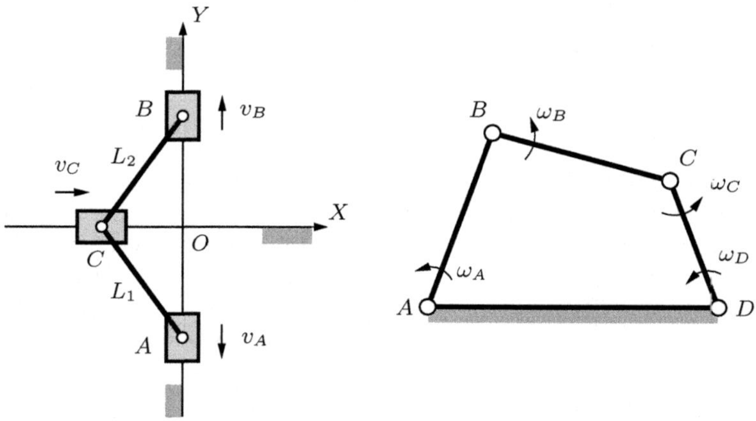

Fig. 2.10 (*Left*) A 1-DOF mechanism with three sliders. The prismatic joints in A and B are on a line perpendicular to the direction of the prismatic joint at C. (*Right*) A 4-bar mechanism. The angular velocities indicated refer to relative motions, e.g. ω_B is the angular velocity of link BC relative to link AB

$$\mathbf{L}_z \begin{bmatrix} \mathbf{m}_v \\ \mathbf{m}_p \end{bmatrix} = \mathbf{0}$$

with $\mathbf{m}_v \neq \mathbf{0}$. Since such a vector exists whenever there exists a unit vector with $\mathbf{m}_v \neq \mathbf{0}$, $\mathbf{q}$ is a singularity of RI type if, and only if, the system of equations

$$\left.\begin{array}{r} \boldsymbol{\Phi}(\mathbf{q}) = \mathbf{0} \\ \mathbf{L}_z\boldsymbol{\xi} = \mathbf{0} \\ \|\boldsymbol{\xi}\|^2 = 1 \end{array}\right\} \quad (2.12)$$

is satisfied for some value of $\boldsymbol{\xi} = (\mathbf{m}_v, \mathbf{m}_p)$ with $\mathbf{m}_v \neq \mathbf{0}$.

Two examples of such singularities are provided in Table 2.1, top-left cell. In the top configuration, the instantaneous input v_A can have any value, while the velocity of point C must be zero and, thus, point B cannot move. Similarly, in the bottom configuration the output link DC cannot move, since the velocity of point C must be zero, while the instantaneous input, ω_A, can have any value.

Redundant Output

A configuration is a singularity of RO type if there exists an output velocity vector $\mathbf{m}_u \neq \mathbf{0}$ and a vector $\mathbf{m}_p$ that satisfy the velocity equation for $\mathbf{m}_v = \mathbf{0}$, i.e. such that

$$\mathbf{L}_y \begin{bmatrix} \mathbf{m}_u \\ \mathbf{m}_p \end{bmatrix} = \mathbf{0}$$

with $\mathbf{m}_u \neq \mathbf{0}$. Following a similar reasoning as above, $\mathbf{q}$ is of RO type if, and only if, it satisfies the equations

$$\left.\begin{array}{r} \boldsymbol{\Phi}(\mathbf{q}) = \mathbf{0} \\ \mathbf{L}_y\boldsymbol{\xi} = \mathbf{0} \\ \|\boldsymbol{\xi}\|^2 = 1 \end{array}\right\} \quad (2.13)$$

for some value of $\boldsymbol{\xi} = (\mathbf{m}_u, \mathbf{m}_p)$ with $\mathbf{m}_u \neq \mathbf{0}$.

The 3-slider and the 4-bar mechanisms in the second column of Table 2.1 are shown in a singularity of RO type. On the former, the instantaneous output v_B can have any value while point A must have zero velocity. The same happens on the latter, where the input link AB is locked while the instantaneous output, ω_D, can have any value.

Impossible Output

A configuration is a singularity of IO type if there exists a vector $\mathbf{m}_u \neq \mathbf{0}$ in the ouput-velocity space for which the velocity equation cannot be satisfied for any combination of $\mathbf{m}_v$ and $\mathbf{m}_p$. In other words, this means that there is a nonzero vector $(\mathbf{m}_u, \mathbf{0}, \mathbf{0})$ that cannot be obtained by projection of any vector $(\mathbf{m}_u, \mathbf{m}_v, \mathbf{m}_p)$ belonging to the kernel of $\mathbf{L}$.

Table 2.1 The six singularity types exemplified with the 3-slider and 4-bar mechanisms

RI, IO	RO, II
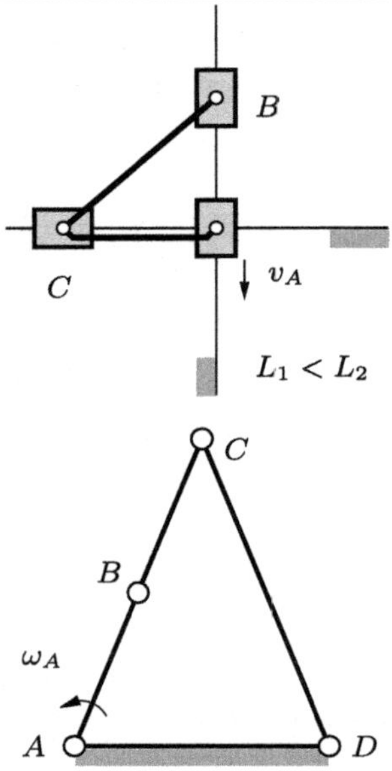	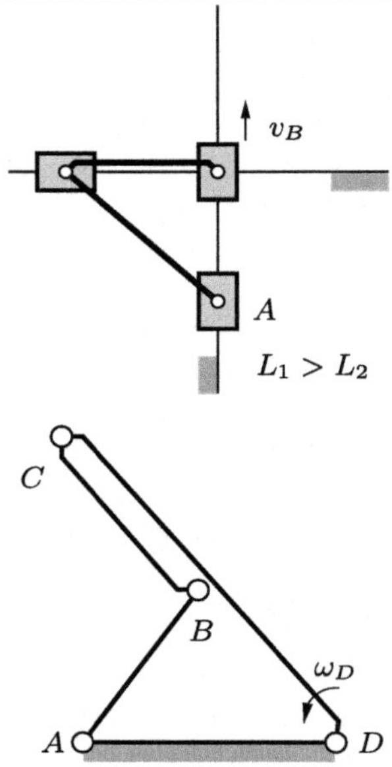

RPM	IIM
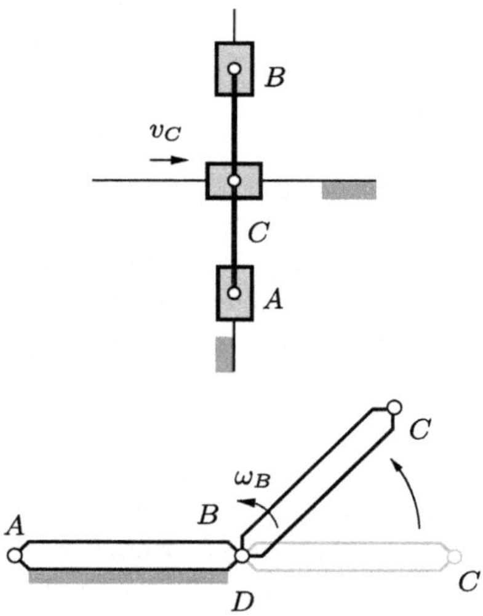	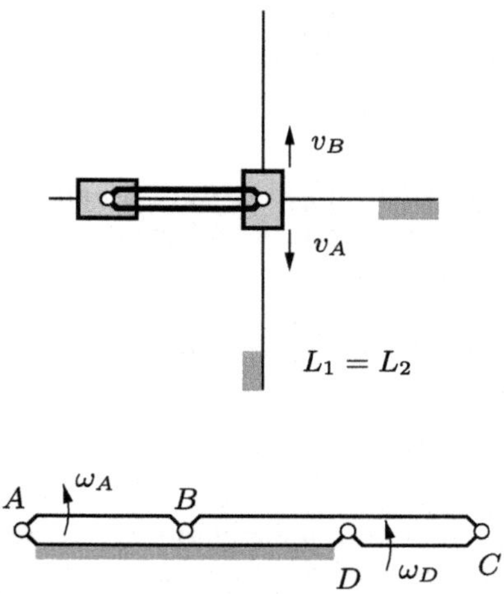

In order to derive the system of equations for this type, let $\mathbf{B} = [\mathbf{b}_1, \ldots, \mathbf{b}_r]$ be a matrix whose columns $\mathbf{b}_i$ form a basis of the kernel of $\mathbf{L}$. Then, the vectors $\mathbf{m}_u$ such that $(\mathbf{m}_u, \mathbf{0}, \mathbf{0})$ can be obtained by projection of some vector of the kernel of $\mathbf{L}$ are those in the column space of

$$\mathbf{A} = \left[\mathbf{I}_{n\times n}\ \mathbf{0}\right]\mathbf{B}.$$

Thus, a singular configuration is of IO type if the map is not surjective, i.e. if $\mathbf{A}$ is rank deficient. In this situation it can be seen that there exists a nonzero vector $\mathbf{m}_u^*$ in the kernel of $\mathbf{A}^\mathsf{T}$ and, hence, a vector $(\mathbf{m}_u^*, \mathbf{0}, \mathbf{0})$ in the kernel of $\mathbf{B}^\mathsf{T}$. Such a vector is orthogonal to all vectors $\mathbf{b}_1, \ldots, \mathbf{b}_r$, so it must belong to the image of $\mathbf{L}^\mathsf{T}$. In conclusion, there must exist a nonzero vector $\mathbf{m}_u^*$ satisfying

$$\mathbf{L}^\mathsf{T}\boldsymbol{\zeta} = \begin{bmatrix} \mathbf{m}_u^* \\ \mathbf{0} \\ \mathbf{0} \end{bmatrix}$$

for some vector $\boldsymbol{\zeta}$, which can be chosen of unit norm. Therefore, a configuration $\mathbf{q}$ is an IO type sigularity if, and only if, it satisfies

$$\left.\begin{aligned} \boldsymbol{\Phi}(\mathbf{q}) &= \mathbf{0} \\ \mathbf{L}^\mathsf{T}\boldsymbol{\zeta} &= \begin{bmatrix} \mathbf{m}_u^* \\ \mathbf{0} \\ \mathbf{0} \end{bmatrix} \\ \|\boldsymbol{\zeta}\|^2 &= 1 \end{aligned}\right\} \tag{2.14}$$

with $\mathbf{m}_u^* \neq \mathbf{0}$. For all solutions of this system, the obtained value of $\mathbf{m}_u^*$ corresponds to a nonfeasible output at the corresponding configuration.

The examples in the first column of Table 2.1 are also singularities of IO type since, in both configurations, any nonzero output is impossible.

Impossible Input

A configuration is a singularity of II type if there exists an input velocity vector $\mathbf{m}_v \neq \mathbf{0}$ for which the velocity equation cannot be satisfied for any combination of $\mathbf{m}_u$ and $\mathbf{m}_p$. Following similar observations as for the IO type, a configuration $\mathbf{q}$ is a singularity of II type if, and only if, there exists a nonzero vector $\mathbf{m}_v^*$ such that

$$\mathbf{L}^\mathsf{T}\boldsymbol{\zeta} = \begin{bmatrix} \mathbf{0} \\ \mathbf{m}_v^* \\ \mathbf{0} \end{bmatrix}$$

for some vector $\boldsymbol{\zeta}$, which can also be chosen of unit norm. Thus, a configuration $\mathbf{q}$ will be a singularity of II type if, and only if, it satisfies

$$\left.\begin{array}{r}\mathbf{\Phi}(\mathbf{q}) = \mathbf{0}\\ \mathbf{L}^{\mathsf{T}}\boldsymbol{\zeta} = \begin{bmatrix}\mathbf{0}\\ \mathbf{m}_v^*\\ \mathbf{0}\end{bmatrix}\\ \|\boldsymbol{\zeta}\|^2 = 1\end{array}\right\} \tag{2.15}$$

with $\mathbf{m}_v^* \neq \mathbf{0}$.

The 3-slider and the 4-bar mechanisms in the second column of Table 2.1 are also singularities of II type since any nonzero input is impossible in these configurations.

Redundant Passive Motion

A configuration is a singularity of RPM type if there exists a passive velocity vector $\mathbf{m}_p \neq \mathbf{0}$ that satisfies the velocity equation for $\mathbf{m}_v = \mathbf{0}$ and $\mathbf{m}_u = \mathbf{0}$, i.e. such that

$$\mathbf{L}_p\mathbf{m}_p = \mathbf{0}$$

with $\mathbf{m}_p \neq \mathbf{0}$. This will happen when the kernel of $\mathbf{L}_p$ is nontrivial and, thus, the following system of equations

$$\left.\begin{array}{r}\mathbf{\Phi}(\mathbf{q}) = \mathbf{0}\\ \mathbf{L}_p\boldsymbol{\xi} = \mathbf{0}\\ \|\boldsymbol{\xi}\|^2 = 1\end{array}\right\} \tag{2.16}$$

encodes all RPM type singularities.

Two examples of these singularities are provided in the first column of Table 2.1. On the 3-slider mechanism, both the input A and the output B must have zero velocity, while the velocity of point C can be nonzero. A 4-bar mechanism with a kite geometry, as shown in the table, can collapse so all joints lie on a single line and B and D coincide. If the input and output are the velocities at joints A and C, ω_A and ω_C, the mechanism can move from the configuration shown in grey, maintaining zero-velocity both at the input and the output joints. Only nonzero velocity is present at the passive joints B and D. Hence, both mechanisms are shown in a singularity of RPM type.

Increased Instantaneous Mobility

A configuration is a singularity of IIM type if $\mathbf{L}$ is rank deficient. In fact, these are the configurations where the instantaneous mobility is greater than the number of degrees of freedom. The definition directly allows us to write the system of equations

$$\left.\begin{array}{r}\mathbf{\Phi}(\mathbf{q}) = \mathbf{0}\\ \mathbf{L}^{\mathsf{T}}\boldsymbol{\xi} = \mathbf{0}\\ \|\boldsymbol{\xi}\|^2 = 1\end{array}\right\}, \tag{2.17}$$

which will be satisfied for some $\boldsymbol{\xi}$ by a configuration $\mathbf{q}$ if, and only if, it is a singularity of IIM type. These correspond to the C-space singularities, where both the FIKP and

IIKP become indeterminate for any choice of input or output among the given velocity variables.

The mobility of the 3-slider and the 4-bar mechanisms in the second column of Table 2.1, bottom, increases from 1 to 2 at the shown configurations and, thus, they are singularities of IIM type.

2.4 A Simple Mechanism with All Singularities

To exemplify how the previous equations can be used to obtain the configurations of each singularity type, consider the 3-slider mechanism in Fig. 2.10. Let (x_P, y_P) denote the coordinates of points $P = A$, B, or C relative to the reference frame OXY in the figure, and let L_1 and L_2 be the lengths of the connector links. Clearly, a configuration of the mechanism can be shortly described by the tuple $\mathbf{q} = (y_A, y_B, x_C)$, because $x_A = x_B = y_C = 0$ in any configuration. Since the distances from A to B and from B to C must be kept equal to L_1 and L_2, Eq. (2.1) is

$$\left.\begin{array}{l} {y_A}^2 + {x_C}^2 = {L_1}^2 \\ {y_B}^2 + {x_C}^2 = {L_2}^2 \end{array}\right\}, \tag{2.18}$$

from which we realise that the C-space corresponds to the intersection of two cylinders in the (y_A, y_B, x_C) space.

The velocity equation in (2.2) could now be obtained using the revolute- and prismatic-joint screws [1], but a more compact expression can here be derived by differentiating Eq. (2.18) with respect to time. Taking v_A and v_B as the input and output velocities, the differentiation yields

$$\mathbf{L} \cdot \mathbf{m} = \begin{bmatrix} 0 & 2y_A & 2x_C \\ 2y_B & 0 & 2x_C \end{bmatrix} \cdot \begin{bmatrix} v_B \\ v_A \\ v_C \end{bmatrix} = \mathbf{0},$$

from where we readily obtain

$$\mathbf{L}_y = \begin{bmatrix} 0 & 2x_C \\ 2y_B & 2x_C \end{bmatrix},$$

$$\mathbf{L}_z = \begin{bmatrix} 2y_A & 2x_C \\ 0 & 2x_C \end{bmatrix},$$

$$\mathbf{L}_p = \begin{bmatrix} 2x_C \\ 2x_C \end{bmatrix},$$

which allow us to define any of the systems in Eqs. (2.7), (2.8) and (2.12)–(2.17).

These systems can be solved analytically in this case. For example, if $L_1 = L_2 = 1$, the C-space has a single connected component composed of two ellipses intersecting on the x_C axis (Fig. 2.11a), and the solutions of the systems in Eqs. (2.7) and (2.8) reveal that the singularity set has six isolated configurations, marked in red in Fig. 2.11a bottom, with the following values of $\mathbf{q}$:

$$\begin{array}{rr} (0,0,1), & (0,0,-1), \\ (1,1,0), & (1,-1,0), \\ (-1,-1,0), & (-1,1,0). \end{array}$$

All of these configurations satisfy both systems, so they are all forward and inverse singularities at the same time. Both the FIKP and the IIKP are indeterminate in them and, therefore, the control of the input or the output velocity does not determine the overall motion of the mechanism.

It turns out, moreover, that the four configurations with $x_C = 0$ satisfy Eqs. (2.14)–(2.16), meaning that they are singularities of IO, II, and RPM type. The other two configurations, which lie on the x_C axis, are singularities of RI, RO, and IIM type because they satisfy Eqs. (2.12), (2.13) and (2.17). These two configurations are in fact C-space singularities. In this case, the C-space self-intersects at these points, and presents a bifurcation that allows the mechanism to change its mode of operation,

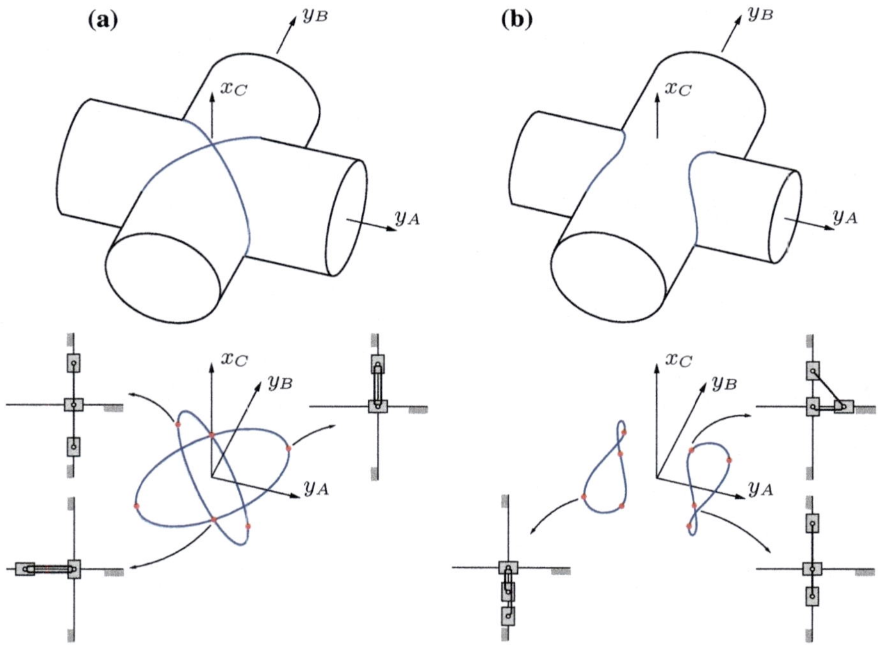

Fig. 2.11 C-space (in *blue*) and singularities (*red dots*) of the 3-slider mechanism for $L_1 = L_2$ (**a**) and $L_1 > L_2$ (**b**) with some examples of singular configurations depicted. In this mechanism, the C-space corresponds to the intersection of two cylinders at right angles

from both sliders moving along the same side of the horizontal axis, $y_A y_B \geq 0$, to one slider moving along each side, $y_A y_B \leq 0$.

The topology of the C-space changes when $L_1 \neq L_2$, since the C-space no longer presents any bifurcation, and it is instead formed by two smooth connected components (Fig. 2.11b). By solving Eqs. (2.7) and (2.8) for $L_1 = 1$ and $L_2 = 0.8$, for example, eight singularities are obtained:

$$\begin{array}{rr} (1, 0.8, 0), & (-1, -0.8, 0), \\ (-1, 0.8, 0), & (0.6, 0, -0.8), \\ (1, -0.8, 0), & (-0.6, 0, 0.8), \\ (0.6, 0, 0.8), & (-0.6, 0, -0.8). \end{array}$$

As before, the configurations with $x_C = 0$ are singularities of IO, II, and RPM type, but the other four configurations are of RO and II type, and there are no singularities of RI or IIM type. To change the operation mode from $y_A \geq 0$ to $y_A \leq 0$ the mechanism has to be disassembled in this case.

It must be noted that if a singularity identification were attempted by means of an input-output velocity equation, for instance by using the equation $y_A v_A = y_B v_B$ which holds for all configurations, the singularities with $x_C = 0$ would not be detected.

References

1. D. Zlatanov, Generalized Singularity Analysis of Mechanisms. PhD thesis, University of Toronto, 1998
2. O. Bohigas, M. Manubens, L. Ros, Singularities of non-redundant manipulators: a short account and a method for their computation in the planar case. Mech. Mach. Theory **68**, 1–17 (2013)
3. J.G. De Jalón, E. Bayo, *Kinematic and Dynamic Simulation of Multibody Systems* (Springer, 1993)
4. S.G. Krantz, H.R. Parks, *The Implicit Function Theorem: History, Theory and Applications* (Birkhäuser, 2002)
5. C.M. Gosselin, J. Angeles, Singularity analysis of closed-loop kinematic chains. IEEE Trans. Robot. Autom. **6**(3), 281–290 (1990)
6. O. Ma, J. Angeles, Architecture singularities of parallel manipulators. Int. J. Robot. Autom. **7**(1), 23–29 (1992)
7. V. Guillemin, A. Pollack, *Differential Topology* (American Mathematical Society, 2010)
8. M. Golubitsky, V. Guillemin, *Stable Mappings and Their Singularities* (Springer, 1973)
9. J. Canny, *The Complexity of Robot Motion Planning* (The MIT Press, Cambridge, 1988)
10. Companion web page of this book. http://www.iri.upc.edu/srm. Accessed 16 Jun 2016
11. I.A. Bonev, Geometric Analysis of Parallel Mechanisms. PhD thesis, Faculté des Sciences et de Génie, Université de Laval, 2002

Chapter 3
Numerical Computation of Singularity Sets

This chapter presents a method to compute any of the singularity sets defined in Chap. 2. The method is very general and applicable to any nonredundant mechanism with lower-pair joints. It allows us to obtain the complete singularity set at the desired accuracy, and to compute each of its singularity subsets independently.

After justifying the approach (Sect. 3.1), we provide a general way of formulating the systems of Eqs. (2.7), (2.8), (2.12)–(2.17), and then present a numerical method able to solve any of such systems (Sect. 3.3). Since the singularity set is often embedded in a highly-dimensional space, it cannot be directly visualised, but we then explain how to project this set in a way convenient to a robot designer (Sect. 3.4). We finally demonstrate the performance of the method and the complexities of the singularity set with the analysis of several robot mechanisms (Sect. 3.5).

3.1 A Suitable Approach

A complete solution of nonlinear systems like those of Eqs. (2.7), (2.8), (2.12)–(2.17) typically involves formulating their equations algebraically, to then take advantage of some suitable symbolic or numerical technique for polynomial system solving. Unfortunately, a good solution to both algebraisation and resolution, treated as independent problems, does not necessarily lead to an efficient solution of the original problem. Our aim has been on finding a good combination of algebraisation and resolution, so that the whole process becomes easy to understand and implement, yet computationally efficient in practice.

Common root-finding approaches for polynomial equations can be classified into algebraic-geometric, continuation, or branch-and-prune techniques. Within the first group of techniques, the most usual ones apply elimination methods to reduce the equations to a minimal set of resultant polynomials, so that the roots of these polynomials, once backsubstituted into intermediate equations, yield all solutions of the

O. Bohigas et al., *Singularities of Robot Mechanisms*,
Mechanisms and Machine Science 41, DOI 10.1007/978-3-319-32922-2_3

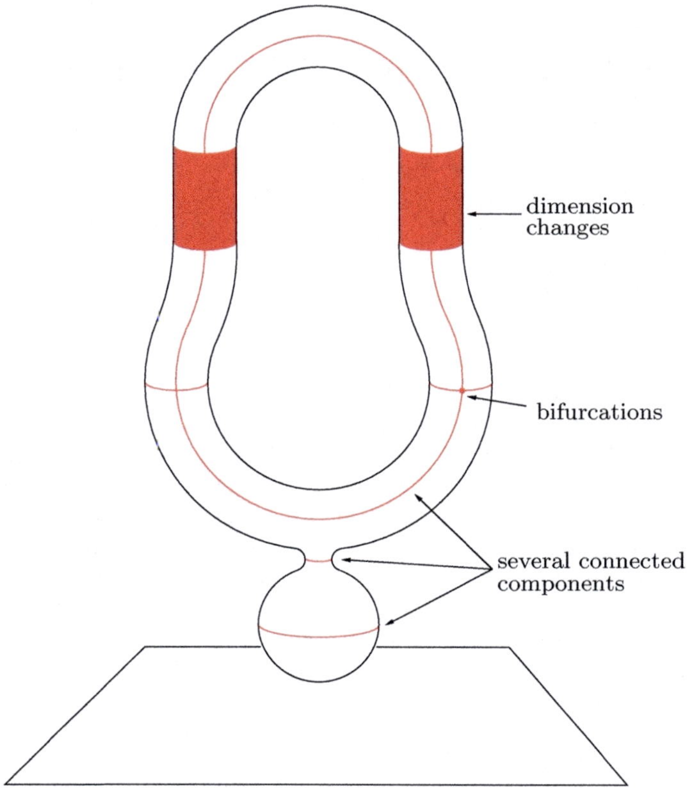

Fig. 3.1 The singularity set is an algebraic variety that may contain bifurcations, dimension changes, or several connected components, even though the C-space is smooth everywhere. The figure shows, in *red*, the singularities that would arise as critical points of the projection map onto the *bottom* plane

original problem [1, 2]. Continuation methods, in contrast, begin with an initial system whose solutions are known, and then transform it gradually to the system whose solutions are sought, while tracking all solution paths along the way [3–6]. Branch-and-prune methods proceed by exclusion instead. They start with an initial box bounding the solutions, and then use relaxed versions of the equations to iteratively eliminate portions of the box that cannot contain a solution. The box is bisected and the process is repeated recursively until a fine-enough approximation of the solution set is obtained [7–9].

To determine the proper kind of approach in our case, we must bear in mind that the structure of the singularity set $\mathcal{S}$ can be very general. Generically, $\mathcal{S}$ has codimension one with respect the C-space, so we may have to deal with singularity sets of dimension 0, 1, 2, or even higher. Elimination and continuation methods are quite general, but they are typically designed to solve systems with isolated or one-dimensional solutions at most. While it is true that continuation methods have been extended to characterise positive-dimensional solutions [6, Part III], they do not directly provide an explicit representation of all solution points in general. Moreover, they work in the complex domain, which adds extra cost when only the real solutions are needed [10], as in our case. Higher-dimensional continuation techniques would in principle be appropriate [11], but they should be fed with a solution point for each connected component of $\mathcal{S}$, and they are still not robust enough to the many nonsmothnesses that $\mathcal{S}$ may eventually present. Note that, even when $\mathcal{C}$ is smooth and contains only one connected component, $\mathcal{S}$ may include multiple connected components, bifurcations, or even dimension changes, among other phenomenona (Fig. 3.1). In contrast, branch-and-prune methods can isolate real solution sets of any dimension and number of connected components, and since they are robust to regularity failures, they are ideal for our purposes. They will be our method of choice in this chapter and the following one.

With regard to the formulation, note that the structure of all systems in Eqs. (2.7), (2.8), (2.12)–(2.17) is very similar. The first line is always Eq. (2.1), because all solution points must correspond to feasible configurations of the mechanism. The second line always involves $\mathbf{L}$ or one of its submatrices, and the third line constrains the norm of some vector. Therefore, the main issue is how to formulate the assembly constraints and the velocity equation in a general way. We next show that, by using a particular choice of configuration coordinates, the formulation of these equations leads to a conceptually-simple branch-and-prune technique for solving any of the Eqs. (2.7), (2.8), (2.12)–(2.17). The formulation closely follows that of reference point coordinates in multibody dynamics [9, 12], which produces simple polynomials with little manipulation, in comparison to other formulations departing from loop closure constraints using relative joint displacements [12, Sect. 2.2.1], or from point-to-point distances [13].

3.2 Formulating the Equations of the Singularity Set

3.2.1 *The Assembly Constraints*

To formulate Eq. (2.1), let us assume that our mechanism has n_b links and n_j joints, labelled $L_1, \ldots, L_{n_b}$, and $J_1, \ldots, J_{n_j}$, respectively, where L_1 is supposed to be the ground link. We furnish every link L_j with a local reference frame, $\mathcal{F}_j$, letting $\mathcal{F}_1$ act as the absolute frame. We shall write $\mathbf{a}^{\mathcal{F}_j}$ to indicate that the components of a vector $\mathbf{a} \in \mathbb{R}^3$ are provided in the basis of $\mathcal{F}_j$, and we shall assume that vectors with no superscript are expressed in the basis of $\mathcal{F}_1$, unless otherwise stated. Then, the pose of each link in the mechanism can be specified by the pair $(\mathbf{r}_j, \mathbf{R}_j)$, where $\mathbf{r}_j \in \mathbb{R}^3$ is the position of the origin of $\mathcal{F}_j$ in frame $\mathcal{F}_1$, and $\mathbf{R}_j \in SO(3)$ is a 3×3 rotation matrix expressing the orientation of $\mathcal{F}_j$ relative to $\mathcal{F}_1$. Note that the link poses cannot be arbitrary though, since they must fulfill the assembly constraints imposed by the joints. We next provide such constraints for the revolute, prismatic, universal, and spherical joints, since these are those typically arising in most mechanisms. However, other lower pairs can be accommodated just as easily using the developments in [9].

If J_i is a revolute joint connecting links L_j and L_k, we consider a point P_i on the axis of the joint, and a unit vector $\mathbf{d}_i$ along the same axis. Then, the assembly constraints of this joint can be written as

$$\mathbf{r}_j + \mathbf{R}_j\, \mathbf{p}_i^{\mathcal{F}_j} = \mathbf{r}_k + \mathbf{R}_k\, \mathbf{p}_i^{\mathcal{F}_k}, \tag{3.1}$$

$$\mathbf{R}_j\, \mathbf{d}_i^{\mathcal{F}_j} = \mathbf{R}_k\, \mathbf{d}_i^{\mathcal{F}_k}, \tag{3.2}$$

where $\mathbf{p}_i$ is the position vector of P_i. The first equation forces the links to be placed so that the point P_i, seen as attached to $\mathcal{F}_j$, coincides with the same point, seen as attached to $\mathcal{F}_k$. The second equation does a similar identification for the $\mathbf{d}_i$ vector (Fig. 3.2a).

If J_i is prismatic, we consider two different points P_i and Q_i on a line parallel to the sliding direction, and a unit vector $\mathbf{d}_i$ pointing in the same direction. In this case, the assembly constraints are equivalent to forcing Q_i to lie on the line of L_j defined by P_i and $\mathbf{d}_i$, while keeping the relative orientation between L_j and L_k fixed to a constant offset (Fig. 3.2b). These conditions are equivalent to

$$\mathbf{r}_j + \mathbf{R}_j\, \mathbf{p}_i^{\mathcal{F}_j} + d_i\, \mathbf{R}_j\, \mathbf{d}_i^{\mathcal{F}_j} = \mathbf{r}_k + \mathbf{R}_k\, \mathbf{q}_i^{\mathcal{F}_k}, \tag{3.3}$$

$$\mathbf{R}_j = \mathbf{R}_{jk} \cdot \mathbf{R}_k, \tag{3.4}$$

where d_i is the linear displacement of the joint and $\mathbf{R}_{jk}$ is a constant rotation matrix providing the relative orientation between links L_j and L_k.

If J_i is universal, the points P_i at the centre of the joint must coincide, and the unit vectors $\mathbf{a}_i$ and $\mathbf{b}_i$ defining the rotation axes must be orthogonal (Fig. 3.2c), which translates into

$$\mathbf{r}_j + \mathbf{R}_j \mathbf{p}_i^{\mathcal{F}_j} = \mathbf{r}_k + \mathbf{R}_k \mathbf{p}_i^{\mathcal{F}_k}, \tag{3.5}$$

$$\mathbf{R}_j \mathbf{a}_i^{\mathcal{F}_j} \cdot \mathbf{R}_k \mathbf{b}_i^{\mathcal{F}_k} = 0. \tag{3.6}$$

If J_i is spherical, the centre points P_i of each link must coincide, while L_k can freely rotate with respect to L_j (Fig. 3.2d). Thus, the valid poses for the two links are those that verify

$$\mathbf{r}_j + \mathbf{R}_j \mathbf{p}_i^{\mathcal{F}_j} = \mathbf{r}_k + \mathbf{R}_k \mathbf{p}_i^{\mathcal{F}_k}. \tag{3.7}$$

Since the coordinates of $\mathbf{p}_i$, $\mathbf{q}_i$, and $\mathbf{d}_i$ in the corresponding local frames are a priori known, the only unknowns appearing in the previous equations are the poses of the links and the displacements of the prismatic joints. The components of the rotation matrices, although unknown, are not independent, since if $\mathbf{R}_j = [\mathbf{s}_j, \mathbf{t}_j, \mathbf{w}_j]$, then it must be

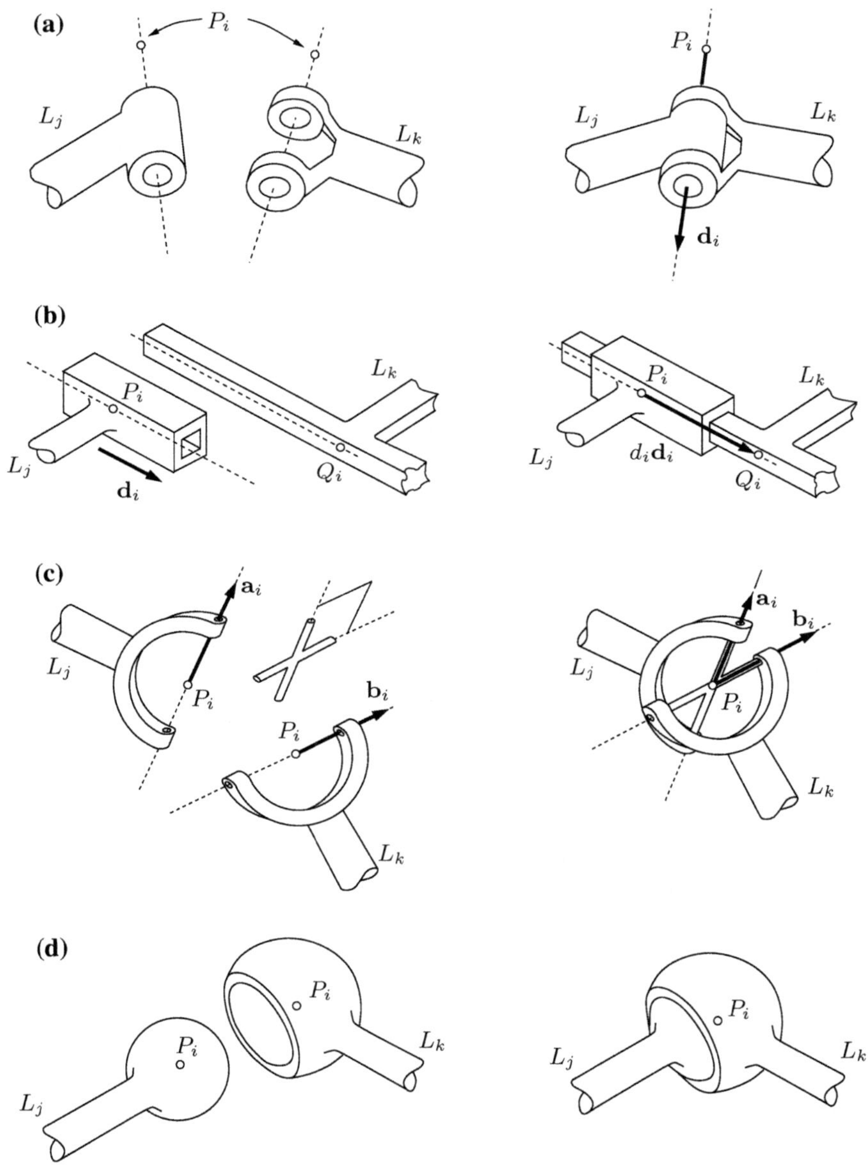

Fig. 3.2 Geometric elements intervening in the assembly of revolute, prismatic, universal, and spherical joints

$$\|\mathbf{s}_j\|^2 = 1, \tag{3.8}$$

$$\|\mathbf{t}_j\|^2 = 1, \tag{3.9}$$

$$\mathbf{s}_j \cdot \mathbf{t}_j = 0, \tag{3.10}$$

$$\mathbf{s}_j \times \mathbf{t}_j = \mathbf{w}_j, \tag{3.11}$$

for $\mathbf{R}_j$ to represent a valid rotation.

In our case, hence, Eq. (2.1) is the system formed by Eqs. (3.1)–(3.11) for all joints and links of the mechanism, and $\mathbf{q}$ is the tuple encompassing the variables $\mathbf{r}_j$ and $\mathbf{R}_j$ of all links, and d_i of all prismatic joints. Since L_1 is the ground link, however, $\mathbf{r}_1 = \mathbf{0}$, and $\mathbf{R}_1$ is the identity matrix.

Note that the resulting system may contain more equations than necessary. For example, Eq. (3.2) introduces a slight redundancy. Since the length of $\mathbf{d}_i$ is known, in some cases it might be sufficient to establish the x and y components of Eq. (3.2). Clearly, the three components are necessary in general, but this is not a nuisance because the methods in Sect. 3.3 can deal with overconstrained systems.

3.2.2 The Velocity Equation

Sometimes, as in planar mechanisms [14], the input and output coordinates directly appear in the assembly constraints, or they can be easily introduced. In these cases, Eq. (2.2) can be obtained by differentiation of Eq. (2.1) as described in Sect. 2.2, but in general it is easier to formulate Eq. (2.2) using the tools of Screw Theory [15, 16]. In favor of generality, thus, we shall formulate Eq. (2.2) using twist loop equations.

It is known that, for a joint J_i between two links, the velocity of one link with respect to the other is given by a linear combination of six-dimensional unit twists [15–17]. Table 3.1 provides the form of these twists and the mentioned linear combination (in axis coordinates) for each one of the joints we consider. As shown, revolute and prismatic joints only involve one unit twist, which is multiplied by the joint velocity ω_i in the linear combination. Instead, universal and spherical joints are equivalent to a combination of two and three revolute joints, and they involve linear combinations of two and three twists, respectively, multiplied by their corresponding joint velocities.

Any set of feasible joint velocities must be in accordance with the kinematic constraints of the mechanism. For that, the sum of the twists intervening along any closed kinematic loop should be zero. If the joint twists in a loop are numbered from 1 to l, this condition can be written as

$$\omega_1 \hat{\mathbf{S}}_1 + \omega_2 \hat{\mathbf{S}}_2 + \cdots + \omega_l \hat{\mathbf{S}}_l = \sum_{i=1}^{l} \omega_i \hat{\mathbf{S}}_i = \mathbf{0}, \tag{3.12}$$

where ω_i is the joint velocity related to the i-th unit twist. This condition must be satisfied for all loops of the mechanism, but only the equations relative to a

Table 3.1 Unit twists and their linear combination for the considered joints

Pair	Shape	Unit twists	Linear combination
Revolute	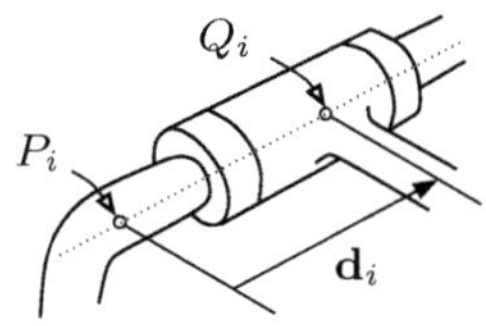	$\hat{\mathbf{S}}_i = \begin{bmatrix} \mathbf{p}_i \times \mathbf{d}_i \\ \mathbf{d}_i \end{bmatrix}$	$\omega_i \hat{\mathbf{S}}_i$
Prismatic	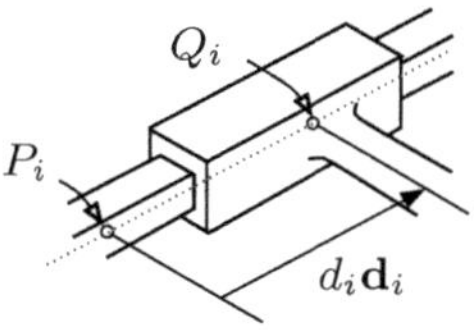	$\hat{\mathbf{S}}_i = \begin{bmatrix} \mathbf{d}_i \\ \mathbf{0} \end{bmatrix}$	$\omega_i \hat{\mathbf{S}}_i$
Universal	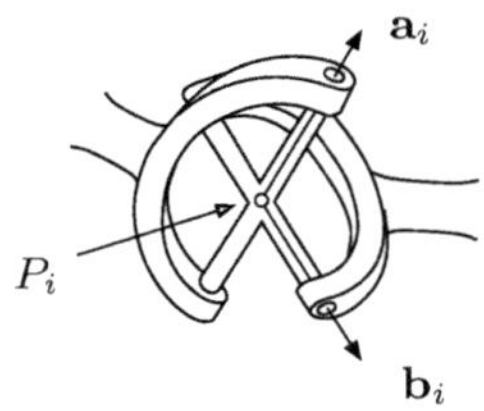	$\hat{\mathbf{S}}_{i,a} = \begin{bmatrix} \mathbf{p}_i \times \mathbf{a}_i \\ \mathbf{a}_i \end{bmatrix}$ $\hat{\mathbf{S}}_{i,b} = \begin{bmatrix} \mathbf{p}_i \times \mathbf{b}_i \\ \mathbf{b}_i \end{bmatrix}$	$\omega_{i,a} \hat{\mathbf{S}}_{i,a} + \omega_{i,b} \hat{\mathbf{S}}_{i,b}$
Spherical	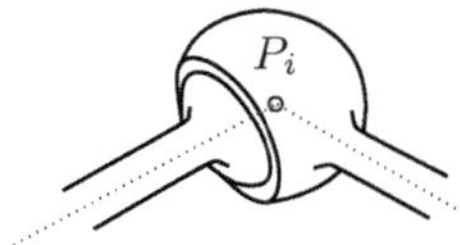	$\hat{\mathbf{S}}_{i,x} = \begin{bmatrix} \mathbf{p}_i \times \mathbf{e}_x \\ \mathbf{e}_x \end{bmatrix}$ $\hat{\mathbf{S}}_{i,y} = \begin{bmatrix} \mathbf{p}_i \times \mathbf{e}_y \\ \mathbf{e}_y \end{bmatrix}$ $\hat{\mathbf{S}}_{i,z} = \begin{bmatrix} \mathbf{p}_i \times \mathbf{e}_z \\ \mathbf{e}_z \end{bmatrix}$	$\omega_{i,x} \hat{\mathbf{S}}_{i,x} +$ $\omega_{i,y} \hat{\mathbf{S}}_{i,y} + \omega_{i,z} \hat{\mathbf{S}}_{i,z}$

Vectors $\mathbf{d}_i$, $\mathbf{a}_i$, and $\mathbf{b}_i$ are those defined in Sect. 3.2.1 for each joint. Vector $\mathbf{p}_i$ provides the position of point P_i of the joint relative to $\mathcal{F}_1$. We can suppose that the spherical joint is instantaneously equivalent to a chain of three revolute joints with axes coincident in P_i, aligned with the vectors $\mathbf{e}_x = (1, 0, 0)$, $\mathbf{e}_y = (0, 1, 0)$, and $\mathbf{e}_z = (0, 0, 1)$

maximal set of independent loops are necessary, since the remaining ones are linearly dependent of them. If the mechanism contains c independent loops, we will have $6c$ equations that impose a necessary and sufficient condition for the feasibility of the joint velocities, including the actuated and passive ones.

In the common case in which each of the output velocity components in $\mathbf{m}_u$ is a component of the twist $\hat{\boldsymbol{T}}$ of the end-effector relative to L_1, such components can be written as a linear function of the joint velocites. If we choose any path of links and joints connecting L_1 to the end-effector and number the joints in this path from 1 to k, we can write

$$\hat{T} = \omega_1 \hat{S}_1 + \cdots + \omega_k \hat{S}_k = \sum_{i=1}^{k} \omega_i \hat{S}_i. \tag{3.13}$$

Therefore, the velocity equation is the system formed by Eq. (3.12) for a set of independent loops, and Eq. (3.13) written for a path connecting the ground link to the end-effector. Identifying terms with Eq. (2.2), it can be seen that $\mathbf{m}$ is the vector containing the joint velocities ω_i and the components of the twist $\hat{T}$, and that $\mathbf{L}$ is the corresponding coefficients matrix. Since the mechanism is nonredundant, if it has N 1-DOF joints and mobility n, then it must have n actuated joints, $N - n$ passive joints, and n output variables, meaning that the size of $\mathbf{m}$ is $N + n$ and, thus, $\mathbf{L}$ is an $N \times (N + n)$ matrix.

The components of $\mathbf{L}$ are given by the unit twists, and since they depend on the configuration $\mathbf{q}$, it will be necessary to add the expressions in the third column of Table 3.1 relating the components of the twists with those of $\mathbf{q}$. Even though these expressions do not include $\mathbf{q}$ explicitly, the vectors $\mathbf{d}_i$, $\mathbf{a}_i$, $\mathbf{b}_i$, and $\mathbf{p}_i$ can be obtained from the components of $\mathbf{q}$ by introducing the following linear relations:

$$\mathbf{d}_i = \mathbf{R}_j \mathbf{d}_i^{\mathcal{F}_j}, \tag{3.14}$$

$$\mathbf{a}_i = \mathbf{R}_j \mathbf{a}_i^{\mathcal{F}_j}, \tag{3.15}$$

$$\mathbf{b}_i = \mathbf{R}_j \mathbf{b}_i^{\mathcal{F}_j}, \tag{3.16}$$

$$\mathbf{p}_i = \mathbf{r}_j + \mathbf{R}_j \mathbf{p}_i^{\mathcal{F}_j}. \tag{3.17}$$

The previous developments allow us to formulate Eqs. (2.7), (2.8), and (2.12)–(2.17) in a way suitable to the method of Sect. 3.3. Each of the systems involve some of the following variables:

- The variables $\mathbf{r}_j$, $\mathbf{R}_j$ of all links, and d_i of all prismatic joints, which constitute the $\mathbf{q}$ tuple.
- The components of the unit twists $\hat{\mathbf{S}}_i$, which yield the components of the $\mathbf{L}$ matrix in Eq. (2.2).
- The vectors $\mathbf{d}_i$, $\mathbf{a}_i$, $\mathbf{b}_i$, and $\mathbf{p}_i$ of Eqs. (3.14)–(3.17).
- The vectors $\boldsymbol{\xi}$ or $\boldsymbol{\zeta}$ needed to establish rank-deficiency conditions.
- The vectors $\mathbf{m}_u^*$ or $\mathbf{m}_v^*$ in the case of IO or II type singularities.

It must be added, however, that the proposed formulation can be usually simplified by eliminating some of the variables and equations involved. For instance, many of the $\mathbf{r}_j$ variables can be eliminated when closed kinematic chains are present in the mechanism, by means of a process explained in detail in [9]. Moreover, in parallel mechanisms it is also possible to use a method to eliminate the passive joint velocities, thus reducing the size of the velocity equation [15] and obtaining more compact systems of equations. Where possible, we make use of these and other simplifications in the computational experiments reported (Sect. 3.5).

3.3 Isolating the Singularity Set

We next provide a method to isolate the solution sets of Eqs. (2.7), (2.8), and (2.12)–(2.17). The method is based on (1) reducing the equations to a simple quadratic form, (2) finding an initial box bounding all of their solution points, and (3) exploiting the quadratic form of the equations to iteratively exclude portions of this box that contain no solution. Such a strategy was proposed in [9, 18] for computing C-spaces of arbitary multiloop linkages, and we have recently applied it successfully to obtain their singularities as well [14, 19]. We shall describe the method steps in detail, analyse its computational cost, and explain its possible parallelization.

3.3.1 Reduction to a Simple Quadratic Form

Observe that in the formulation of Sect. 3.2.2 all of the equations are polynomial, and they only involve linear, bilinear, or quadratic terms. In other words, if x_i and x_j refer to any two of their variables, the monomials can only be of the form x_i, $x_i x_j$, or x_i^2. Let us then define the changes of variables

$$p_k = x_i^2, \tag{3.18}$$

$$b_k = x_i x_j, \tag{3.19}$$

for each quadratic and bilinear monomial, respectively. These changes allow us to convert any of the systems into the expanded form:

$$\left.\begin{array}{l} \boldsymbol{\Lambda}(\mathbf{x}) = 0 \\ \boldsymbol{\Omega}(\mathbf{x}) = 0 \end{array}\right\}, \tag{3.20}$$

where $\mathbf{x}$ is a tuple encompassing the original and the newly-defined variables, $\boldsymbol{\Lambda}(\mathbf{x}) = 0$ is a subsystem of linear equations in $\mathbf{x}$, and $\boldsymbol{\Omega}(\mathbf{x}) = 0$ is a subsystem of quadratic equations of the form of (3.18) or (3.19). Equation (3.20) involves more equations and variables than the original system, but the simpler structure of its equations is beneficial to the branch-and-prune strategy defined in Sect. 3.3.3.

In the systems of Eqs. (2.12)–(2.15), there is a vector that must be different from zero, but since the technique can also handle nonstrict inequalities, as explained below, the latter condition can be enforced by setting

$$\|\mathbf{m}_v\|^2 \geq \epsilon \tag{3.21}$$

for the systems in Eqs. (2.12) and (2.15), and

$$\|\mathbf{m}_u\|^2 \geq \epsilon, \tag{3.22}$$

for the systems in Eqs. (2.13) and (2.14), where ϵ is a sufficiently small value. By using these inequalities, whose terms are also quadratic, some solutions of these systems might be neglected, but ϵ can be made arbitrarily small, reducing the set of missed solutions to a negligible size. Note that this does not prevent us from capturing all singularities, as we can always compute them exhaustively by solving Eqs. (2.7) and (2.8), which only involve equalities.

3.3.2 Initial Bounding Box

An advantage of the proposed formulation is that it is straightforward to define conservative interval bounds for all solutions of Eq. (3.20), since:

- Simple feasibility intervals for the components of $\mathbf{r}_j$ can be derived from the link dimensions.
- The allowed displacement d_i of the prismatic joints is usually restricted by design, or it can be derived from the link dimensions.
- The columns of $\mathbf{R}_j$ are restricted to be unit vectors by Eqs. (3.8)–(3.11), so their components can only take values in the $[-1, 1]$ interval.
- The components of the unit twists $\hat{\mathbf{S}}_i$ can be derived from the intervals of other variables using the expressions of Table 3.1.
- The variables that refer to the components of the unit vectors $\mathbf{d}_i$, $\mathbf{a}_i$, and $\mathbf{b}_i$ can only take values in the $[-1, 1]$ interval.
- The interval for the components of $\mathbf{p}_i$ can be derived using Eq. (3.17).
- The components of $\boldsymbol{\xi}$ and $\boldsymbol{\zeta}$ are bounded in the range $[-1, 1]$ because they are unit vectors.
- The feasibility intervals for the components of $\mathbf{m}_u^*$ and $\mathbf{m}_v^*$ in Eqs. (2.14) and (2.15) can be obtained by mapping known intervals using $\mathbf{L}_u^{\mathsf{T}}\boldsymbol{\zeta} = \mathbf{m}_u^*$ and $\mathbf{L}_v^{\mathsf{T}}\boldsymbol{\zeta} = \mathbf{m}_v^*$.
- Intervals for the p_k and b_k variables can be derived by simple interval operations using Eqs. (3.18) and (3.19).

In conclusion, from the Cartesian product of all these intervals it is possible to define an initial rectangular box $\mathcal{B}$ bounding all solutions of Eq. (3.20).

3.3.3 A Branch-and-Prune Method

The algorithm for solving Eq. (3.20) recursively applies two operations on $\mathcal{B}$: box *shrinking* and box *splitting*. Using box shrinking, portions of $\mathcal{B}$ containing no solution are eliminated by narrowing some of its defining intervals. This process is repeated until either the box is reduced to an empty set, in which case it contains no solution, or the box is "sufficiently" small, in which case it is considered a *solution* box, or the box cannot be "significantly" reduced, in which case it is bisected into two sub-boxes via box splitting (which simply bisects its largest interval). To converge to all solutions, the whole process is recursively applied to the new sub-boxes, until one obtains a collection of solution boxes whose side lengths are below a given threshold σ.

The crucial operation in this scheme is box shrinking, which is implemented as follows. Note first that the solutions falling in some sub-box $\mathcal{B}_c \subseteq \mathcal{B}$ must lie in the linear variety defined by $\boldsymbol{\Lambda}(\mathbf{x}) = 0$. Thus, we may shrink $\mathcal{B}_c$ to the smallest possible box bounding this variety inside $\mathcal{B}_c$. The limits of the shrunk box along dimension x_i can be found by solving the following two linear programs:

$$\textbf{LP1:} \text{ Minimise } x_i \text{, subject to: } \boldsymbol{\Lambda}(\mathbf{x}) = 0, \mathbf{x} \in \mathcal{B}_c,$$
$$\textbf{LP2:} \text{ Maximise } x_i \text{, subject to: } \boldsymbol{\Lambda}(\mathbf{x}) = 0, \mathbf{x} \in \mathcal{B}_c.$$

However, $\mathcal{B}_c$ can be further reduced because the solutions must also satisfy all equations $p_k = x_i^2$ and $b_k = x_i x_j$ in $\boldsymbol{\Omega}(\mathbf{x}) = 0$. These equations can be taken into account by noting that, if $[g_i, h_i]$ denotes the interval of $\mathcal{B}_c$ along dimension x_i, then:

- The portion of the parabola $p_k = x_i^2$ lying inside $\mathcal{B}_c$ is bound by the triangle $A_1 A_2 A_3$, where A_1 and A_2 are the points where the parabola intercepts the lines $x_i = g_i$ and $x_i = h_i$, and A_3 is the point where the tangent lines at A_1 and A_2 meet (Fig. 3.3a).
- The portion of the hyperbolic paraboloid $b_k = x_i x_j$ lying inside $\mathcal{B}_c$ is bound by the tetrahedron $B_1 B_2 B_3 B_4$, where the points $B_1, \ldots, B_4$ are obtained by lifting the corners of the rectangle $[g_i, h_i] \times [g_j, h_j]$ vertically to the paraboloid (Fig. 3.3b).

Linear inequalities corresponding to these bounds can be added to LP1 and LP2, which usually produces a much larger reduction of $\mathcal{B}_c$, or even its complete elimination if one of the linear programs is found unfeasible. In this step, the inequalities needed to model the conditions in Eqs. (3.21) and (3.22) can also be taken into account by adding them to the linear programs.

Algorithm 3.1 gives the top-level pseudocode of the process. As input, it receives the box $\mathcal{B}$, the list $\mathbf{F}$ containing the equations in Eq. (3.20) describing the singularity set to compute, and two threshold parameters σ and ρ. As output, it returns a list $\mathbf{B}$ of

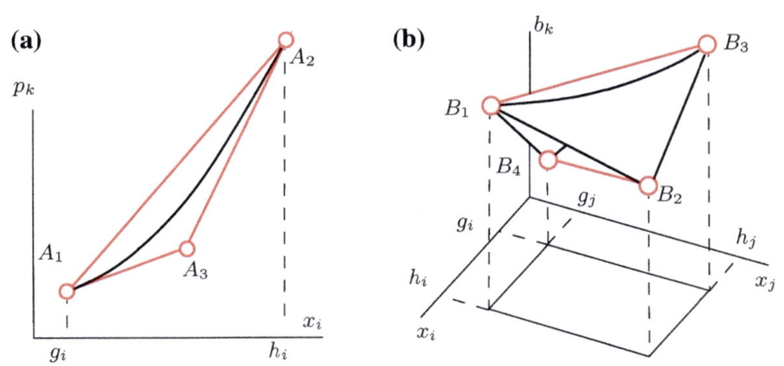

Fig. 3.3 Polytope bounds within box $\mathcal{B}_c$

Algorithm 3.1: Quadratic equation solver.

Solve-system($\mathcal{B}$, **F**, σ, ρ)
input : The initial bounding box, $\mathcal{B}$, a list **F** with the functions defining Eq. (3.20), and the parameters σ and ρ.
output: A list **B** of solution boxes.

1 $\mathbf{B} \leftarrow \varnothing$
2 $\mathbf{P} \leftarrow \{\mathcal{B}\}$
3 **while** $\mathbf{P} \neq \varnothing$ **do**
4     $\mathcal{B}_c \leftarrow$ EXTRACT(**P**)
5   **while** ***not*** IS-VOID($\mathcal{B}_c$) ***and*** SIZE($\mathcal{B}_c$) $> \sigma$ ***and*** $\frac{V_c}{V_p} \leq \rho$ **do**
6         $V_p \leftarrow$ VOLUME($\mathcal{B}_c$)
7   SHRINK-BOX($\mathcal{B}_c$, **F**)
8         $V_c \leftarrow$ VOLUME($\mathcal{B}_c$)
9   **if** ***not*** IS-VOID($\mathcal{B}_c$) **then**
10   **if** SIZE($\mathcal{B}_c$) $\leq \sigma$ **then**
11             $\mathbf{B} \leftarrow \mathbf{B} \cup \{\mathcal{B}_c\}$
12   **else**
13             $(\mathcal{B}_1, \mathcal{B}_2) \leftarrow$ SPLIT-BOX($\mathcal{B}_c$)
14             $\mathbf{P} \leftarrow \mathbf{P} \cup \{\mathcal{B}_1, \mathcal{B}_2\}$
15 RETURN(**B**)

"solution boxes" that enclose all points of the solution set. The functions VOLUME($\mathcal{B}$) and SIZE($\mathcal{B}$) compute the volume and the length of the longest side of $\mathcal{B}$, respectively. These and other low-level procedures of straightforward implementation are left unspecified.

Initially, two lists are set up in lines 1 and 2: an empty list **B** of solution boxes, and a list **P** of boxes to be processed, containing $\mathcal{B}$. A **while** loop is then executed, until **P** gets empty (lines 3–14), by iterating the following steps. Line 4 extracts one box from **P**. Lines 5–8 repeatedly reduce this box as much as possible via the SHRINK- BOX function, until either the box is an empty set (IS- VOID($\mathcal{B}_c$) is true), or it cannot be significantly reduced ($V_c/V_p > \rho$), or it becomes small enough (SIZE($\mathcal{B}$) $\leq \sigma$). In the latter case, the box is considered a solution of the problem. If a box is neither a solution nor an empty box, lines 13 and 14 split it into two sub-boxes and add them to **P** for further processing. The SHRINK- BOX procedure in line 7 takes as input the box $\mathcal{B}_c$ to be shrunken and the list of equations **F**. The procedure collects all linear equations and all half planes approximating the equations $p_k = x_i^2$ and $b_k = x_i x_j$, and then uses these constraints to reduce every dimension of the box by solving the linear programs LP1 ands LP2 indicated above, which possibly provide tighter bounds for the corresponding intervals.

As it turns out, the previous algorithm explores a binary tree of boxes whose internal nodes correspond to boxes that have been split at some time, and whose leaves are either solution or empty boxes. The collection of all solution boxes, which is returned as output in line 15, is said to form a *box approximation* of the solution set of Eq. (3.20), because its boxes constitute a discrete envelope of the set whose accuracy can be adjusted through the σ parameter. Figure 3.4 illustrates the progression of the algorithm and the resulting box approximations when computing the lemniscate curve of Gerono, and the C-space of the 3-slider mechanism of Sect. 2.4.

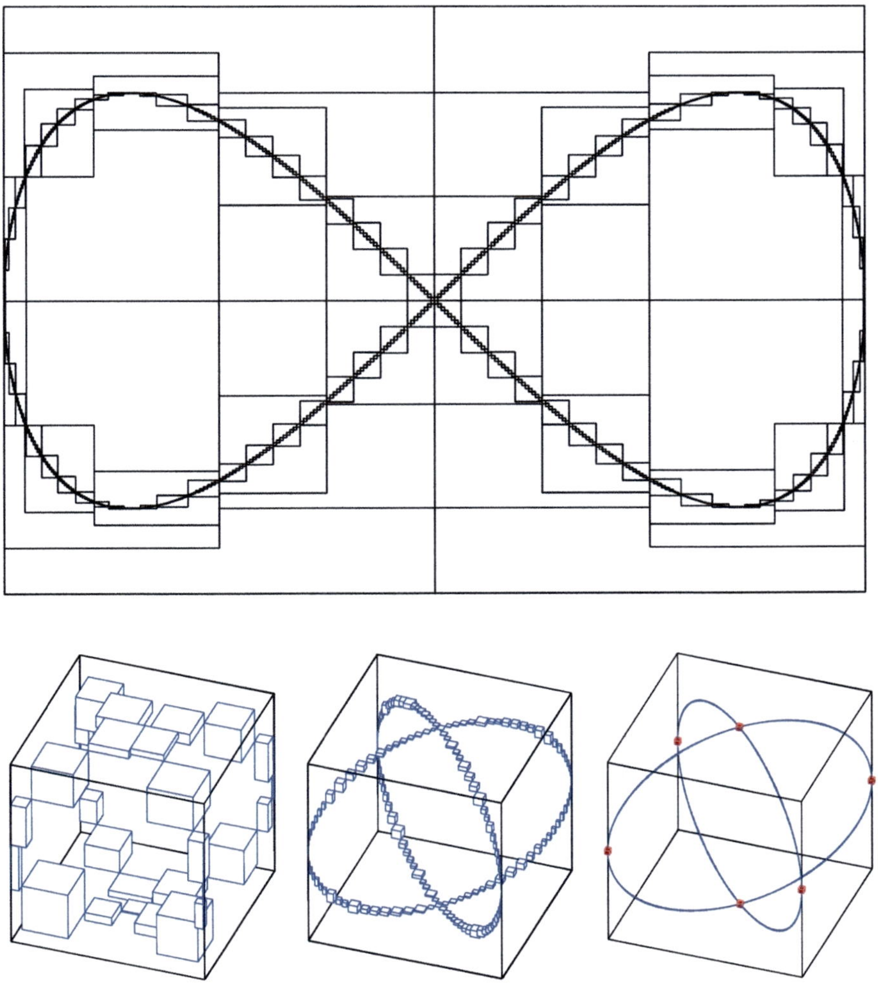

Fig. 3.4 (*Top*) progression of the algorithm when computing the lemniscate curve of Gerono defined by $y^4 = (y^2 - x^2)$. We show the initial box and, overlaid, the intermediate and final box approximations generated by the algorithm. (*Bottom*) progression when computing the C-space of the 3-slider mechanism of Sect. 2.4, i.e., the solution of Eq. (2.18) with $L_1 = L_2$. From *left* to *right*, the sequence shows three stages of the computation, in *blue*, with the singularities of the mechanism shown overlaid in the last plot, in *red*, obtained by solving Eqs. (2.7) and (2.8)

3.3.4 Computational Cost and Parallelization

The computational cost of the algorithm can be evaluated by analysing the cost of one iteration, and the number of iterations that need to be performed, both in terms of the number of bodies (n_b) and joints (n_j) of the mechanism. On the one hand, we can consider that an iteration consists of applying the box shrinking process to a given box. This involves solving $2n_x$ linear programs, where n_x is the number of variables in Eq. (3.20). Since n_x depends linearly on n_b and n_j, and Karmarkar's bound for the complexity of linear programming is $O(n_x^{3.5})$ [20], we can conclude that the cost of one iteration is worst-case polynomial with n_b and n_j. On the other hand, it is difficult to predict how many steps will the algorithm require to isolate all solutions. The number of iterations largely depends on the chosen σ, and on the dimension d of the singularity subset considered. For $d = 0$ the algorithm is relatively fast because it is quadratically convergent to the roots [9]. For $d \geq 1$, the cost is inversely proportional to σ in the best case. For a fixed σ, however, the amount of solution boxes grows exponentially with d, so that an initial guess of the execution time is usually made on the basis of d only. The value of d can be estimated by noting that the singularity set is typically of codimension one relative to the C-space, and using the Grübler-Kutzbach formula with n_b and n_j to determine the C-space dimension. Besides, notice that the algorithm is complete, in the sense that it will succeed in isolating all solution points of Eq. (3.20) accurately, provided that a small-enough value for σ is used. Detailed properties of the algorithm, including an analysis of its completeness, correctness, and convergence order, can be found in [9].

Clearly, the algorithm is computationally demanding, but note that it can be naturally parallelized to be run on multiprocessor computers. To this end, we can just implement the book-keeping of the search tree in a selected "supervisor" processor which keeps track of the tree leafs at all times. Every leaf that is neither an empty nor a solution box needs to be further reduced. Since box reduction is the most time-consuming task, and several boxes await for it simultaneously, it makes sense to perform the reductions in parallel, by assigning each of them to any of the remaining "child" processors. A child processor's task is thus to receive a box from the supervisor, to reduce it as much as possible by solving the aforementioned linear programs, and to return the reduced box back to the supervisor, which will queue it for further splitting and reduction, if needed, or mark it as a solution or an empty box.

3.4 Visualising the Singularity Sets

Even though we have a means to compute $\mathcal{S}$, a nontrivial issue is how to represent this set in a way suitable to the needs of a robot designer. Because of the high number of configuration variables typically involved in $\mathbf{q}$, $\mathcal{S}$ is often embedded in a highly-dimensional space, and the use of 2- or 3-D projections becomes inevitable to understand its structure. An enlightening choice, as done for instance in [21–24], is to project $\mathcal{S}$ to the output space $\mathcal{U}$, since this space encodes the end-effector motion and is easier to interpret. In such a projection, points corresponding to inverse singularities indicate a loss of instantaneous degrees of freedom relative to the **u** variables. As shown in Chap. 4, these points provide the boundary and interior barriers of the workspace relative to such variables. Similarly, $\mathcal{S}$ can be projected to the input space $\mathcal{V}$, as done in [25–27], where the forward singularities delimit the motion range that should be reachable by the actuators. Both the $\mathcal{V}$ and $\mathcal{U}$ spaces get partitioned into several regions after such projections, and it is possible to decide which regions correspond to feasible configurations of the mechanism by selecting a point in each region and solving

$$\mathbf{\Phi}(\mathbf{v}, \mathbf{y}) = \mathbf{0}$$

or

$$\mathbf{\Phi}(\mathbf{u}, \mathbf{z}) = \mathbf{0}$$

with **v** or **u** fixed to the selected point, using the numerical method described in Sect. 3.3.

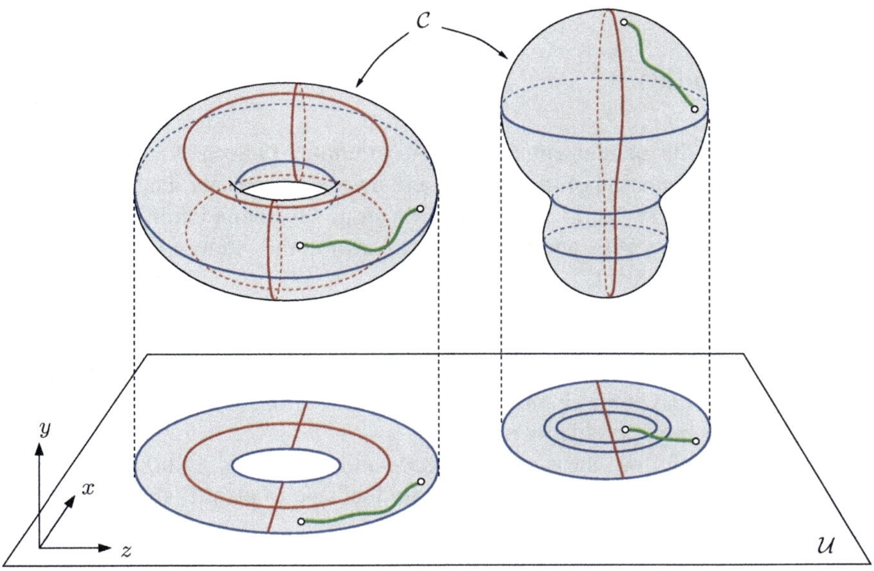

Fig. 3.5 A portrait of a fictitious C-space $\mathcal{C}$ with two connected components. The $\mathcal{V}$ and $\mathcal{U}$ spaces are assumed to be the (x, y) and (x, z) planes in this case, so that the forward and inverse singularity loci are the *red* and *blue curves*, respectively. Only the portrait on the $\mathcal{U}$ space is shown for simplicity. The portrait, as in this case, may reveal the existence of several connected components in $\mathcal{C}$. Also, it can be used as a safe navigation map, because paths in the portrait not crossing a projected singularity correspond to singularity-free paths in $\mathcal{C}$ (*left* path). However, the converse is not necessarily true (*right* path)

The resulting diagrams, which we refer to as singularity *portraits*, convey much global information about the motion capabilities of the mechanism because (Fig. 3.5):

- The existence of several connected components in $\mathcal{C}$ may be revealed by the portrait, and such knowledge may be useful to determine the most appropriate component into which the mechanism should be "assembled" by design, depending on the task to be performed.
- A feasible path in $\mathcal{V}$ or $\mathcal{U}$ not crossing a projected singularity always corresponds to a singularity-free path in $\mathcal{C}$.
- Only when approaching a projected singularity some kind of motion degeneracy is to be expected, so that a portrait can be used as a safe navigation map of $\mathcal{C}$.

Despite these advantages, note that the connectivity of the singularity-free regions of $\mathcal{C}$ is only partially reflected in the portraits. For example, on the right component of Fig. 3.5 it is easy to see that distinct points of $\mathcal{C}$ may seem to be separated by singularities when looking at the portrait, while they are actually connected by singularity-free paths in $\mathcal{C}$. Chapters 5 and 6 will develop robust numerical tools to determine the existence of such paths, and to provide the whole singularity-free region of $\mathcal{C}$ that is reachable from a given configuration.

3.5 Case Studies

We next illustrate the method with a number of examples. In Sects. 3.5.1 and 3.5.2 we apply it to the computation of the forward and inverse singularity sets of planar and spatial mechanisms. In Sect. 3.5.3 we then illustrate the computation of Zlatanov's lower-level singularity sets, and the complex partition they induce on the C-space, using a 2-DOF mechanism.

All computations have been carried out using the parallelized version of the method outlined in Sect. 3.3.3, implemented in C using the libraries of the CUIK Suite [28], executed on a grid computer of Xeon processors able to run 160 threads in parallel. At the end of the chapter, a table summarises the performance data of the method on the example problems reported.

In all plots that follow, the same color code adopted in Fig. 3.5 has been used to distinguish the forward and inverse singularity loci, and to identify the regions of $\mathcal{U}$ and $\mathcal{V}$ attainable by the mechanism.

3.5.1 *The 3-RRR Mechanism*

The 3-RRR mechanism consists of a moving platform connected to ground by means of three legs, where each leg is a three-revolute chain (Fig. 3.6). The most common versions of this mechanism mount the actuators on the intermediate or base joints (Fig. 3.7). Here, we initially consider the 3-R$\underline{\text{R}}$R version, in which the three interme-

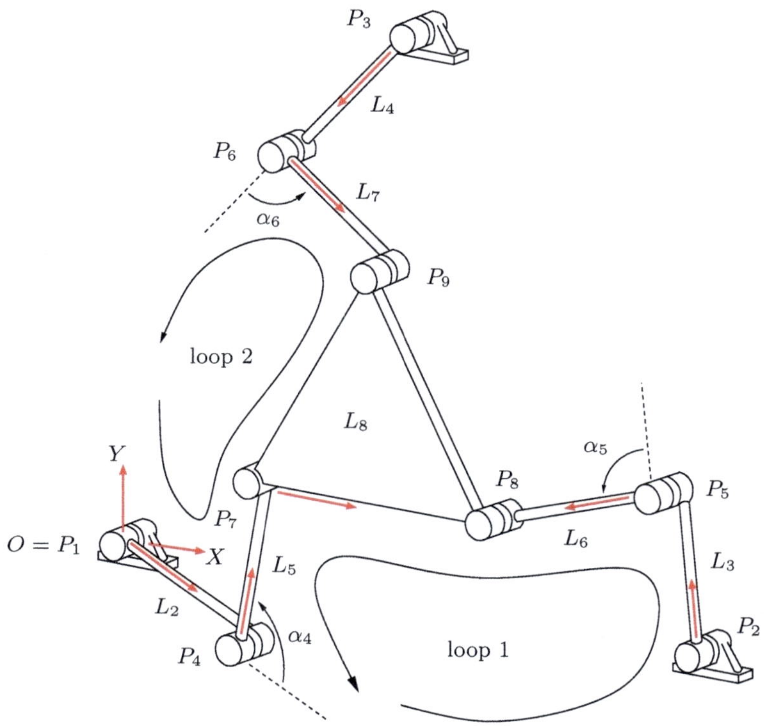

Fig. 3.6 A planar 3-RRR mechanism. Each point P_i is the center of a revolute joint J_i. The links are labelled $L_1, \ldots, L_8$, where L_1 is the ground link $P_1 P_2 P_3$, and L_8 is the end effector $P_7 P_8 P_9$. Every link L_j has an associated reference frame $\mathcal{F}_j$ centred in P_{j-1}, whose X axis is indicated in *red*. The relative angles at the intermediate joints J_i of each leg are labelled as α_i. The absolute angles between frames $\mathcal{F}_j$ and $\mathcal{F}_1$ (not drawn) are denoted by θ_j

diate joints placed at points P_4, P_5, and P_6 are actuated (allowing the independent control of the three degrees of freedom of the platform) and the remaining joints are passive. The input velocities of the mechanism are then the angular velocities at the intermediate joints, and the output ones are given by the twist of the moving platform $P_7 P_8 P_9$, which acts as the end-effector.

Several tools can be used to study the singularity set $\mathcal{S}$ of this mechanism [30–32], which is two-dimensional in general. A good reference summarising them all is [22], where it is shown that the forward singularities can be derived from those of the 3-R$\underline{\text{P}}$R mechanism [30], whereas the inverse singularities can be generated geometrically, from the so-called vertex-spaces of the legs. These methods are useful, but concentrate on deriving the constant-orientation slices of $\mathcal{S}$ only, so that a reconstruction of the whole singularity surface involves a discretisation of the orientation angle of the platform, which necessarily leaves points of $\mathcal{S}$ out of the representation. Moreover, only projections of the slices to the plane of two output variables are

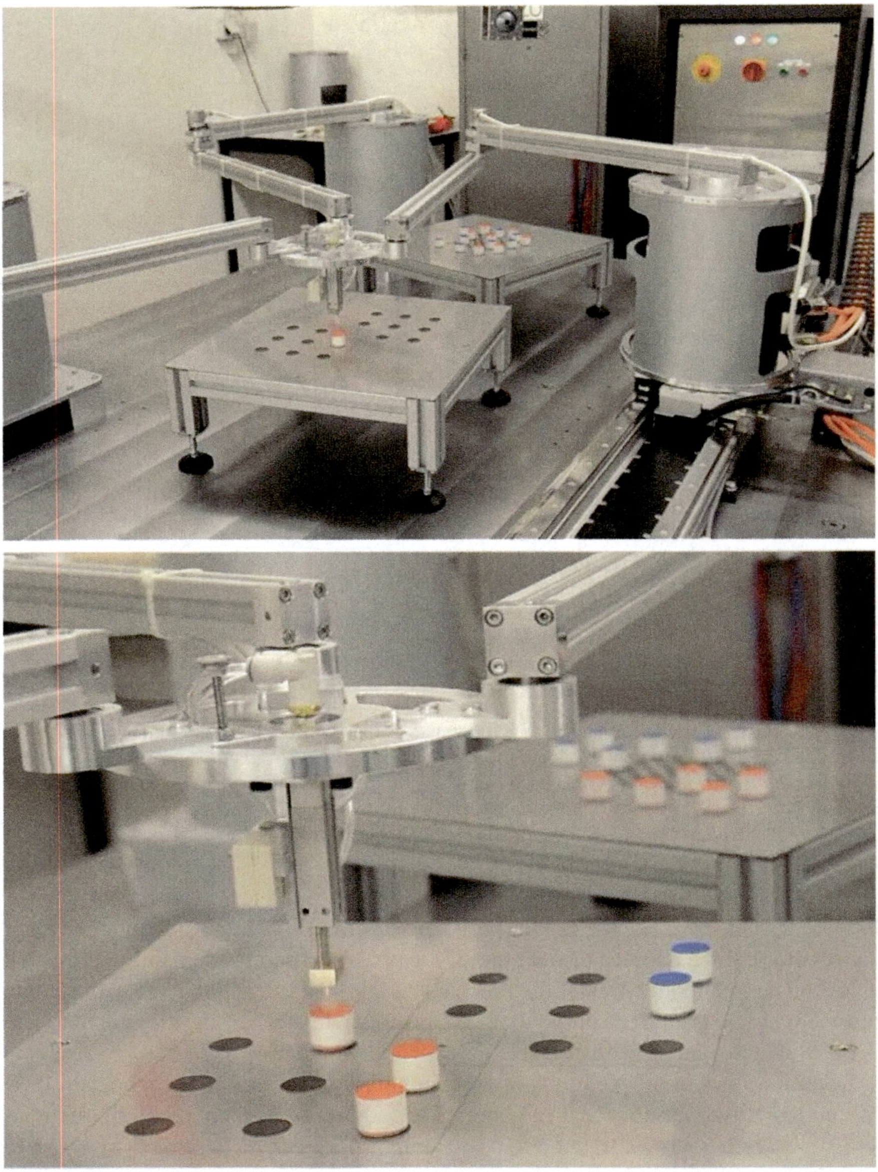

Fig. 3.7 A reconfigurable planar $3\text{-}\underline{R}RR$ mechanism with actuators on the base revolute joints, built at the Leibniz Universität Hannover [29]. The robot is shown in a fast pick-and-place scenario. One of the actuated joints can be placed at will along a slider, in order to fix an architecture that avoids the crossing of singularities during the task (Courtesy of Dr. Jens Kotlarski.)

derived, implying that the visualisation of the singularity surface on the input space, for example, is not straightforward. The method presented here, in contrast, allows us to compute the whole singularity surface directly in $\mathcal{C}$, and to project it easily to any required space, including $\mathcal{V}$ or $\mathcal{U}$, without incurring in any loss of information.

To compute $\mathcal{S}$, the proposed method requires formulating the assembly constraints as explained in Sect. 3.2. For this purpose, we label the links of the mechanism as $L_1, \ldots, L_8$, where L_1 is the ground link defined by P_1, P_2, and P_3, L_8 is the moving platform $P_7 P_8 P_9$, and the other links L_j are those defined by points P_{j-1} and P_{j+2}. We also label the joints as $J_1, \ldots, J_8$, where J_i is the joint at point P_i. For the particular mechanisms under study, the coordinates of points P_i in the local frames of each link are given in Table 3.2, which correspond to the geometries analysed in [22, Sect. 2.4].

Since this is a planar mechanism, the position vectors $\mathbf{r}_j$ of the links provide points in $\mathbb{R}^2$, and the matrices $\mathbf{R}_j$ can be written as

$$\mathbf{R}_j = \begin{bmatrix} c_j & -s_j \\ s_j & c_j \end{bmatrix},$$

where c_j and s_j are the cosine and sine of the angle θ_j between $\mathcal{F}_j$ and $\mathcal{F}_1$, and thus verify

$$c_j^2 + s_j^2 = 1. \tag{3.23}$$

Note that the assembly constraint of a revolute joint in the plane is equivalent to imposing Eq. (3.1) only. Therefore, Eq. (2.1) is in this case the system formed by Eq. (3.1) for all joints, and Eq. (3.23) for all links. However, as shown in [9], it is possible to eliminate the $\mathbf{r}_j$ variables by replacing all instances of Eqs. (3.1) by their

Table 3.2 Joint coordinates of the two 3-RRR mechanisms considered

Robot	L_1	L_2	L_3	L_4
3-R$\underline{\text{R}}$R	$\mathbf{p}_1^{\mathcal{F}_1} = (0, 0)$ $\mathbf{p}_2^{\mathcal{F}_1} = (-2.386, 0)$ $\mathbf{p}_3^{\mathcal{F}_1} = (-1.193, -2.067)$	$\mathbf{p}_1^{\mathcal{F}_2} = (0, 0)$ $\mathbf{p}_4^{\mathcal{F}_2} = (4, 0)$	$\mathbf{p}_2^{\mathcal{F}_3} = (0, 0)$ $\mathbf{p}_5^{\mathcal{F}_3} = (4, 0)$	$\mathbf{p}_3^{\mathcal{F}_4} = (0, 0)$ $\mathbf{p}_6^{\mathcal{F}_4} = (4, 0)$
3-$\underline{\text{R}}$RR	$\mathbf{p}_1^{\mathcal{F}_1} = (0, 0)$ $\mathbf{p}_2^{\mathcal{F}_1} = (2.35, 0)$ $\mathbf{p}_3^{\mathcal{F}_1} = (1.175, 2.035)$	$\mathbf{p}_1^{\mathcal{F}_2} = (0, 0)$ $\mathbf{p}_4^{\mathcal{F}_2} = (1, 0)$	$\mathbf{p}_2^{\mathcal{F}_3} = (0, 0)$ $\mathbf{p}_5^{\mathcal{F}_3} = (1, 0)$	$\mathbf{p}_3^{\mathcal{F}_4} = (0, 0)$ $\mathbf{p}_6^{\mathcal{F}_4} = (1, 0)$
Robot	L_5	L_6	L_7	L_8
3-R$\underline{\text{R}}$R	$\mathbf{p}_4^{\mathcal{F}_5} = (0, 0)$ $\mathbf{p}_7^{\mathcal{F}_5} = (3, 0)$	$\mathbf{p}_5^{\mathcal{F}_6} = (0, 0)$ $\mathbf{p}_8^{\mathcal{F}_6} = (3, 0)$	$\mathbf{p}_6^{\mathcal{F}_7} = (0, 0)$ $\mathbf{p}_9^{\mathcal{F}_7} = (3, 0)$	$\mathbf{p}_7^{\mathcal{F}_8} = (0, 0)$ $\mathbf{p}_8^{\mathcal{F}_8} = (-0.276, 0.276)$ $\mathbf{p}_9^{\mathcal{F}_8} = (-0.919, 0.184)$
3-$\underline{\text{R}}$RR	$\mathbf{p}_4^{\mathcal{F}_5} = (0, 0)$ $\mathbf{p}_7^{\mathcal{F}_5} = (1.35, 0)$	$\mathbf{p}_5^{\mathcal{F}_6} = (0, 0)$ $\mathbf{p}_8^{\mathcal{F}_6} = (1.35, 0)$	$\mathbf{p}_6^{\mathcal{F}_7} = (0, 0)$ $\mathbf{p}_9^{\mathcal{F}_7} = (1.35, 0)$	$\mathbf{p}_7^{\mathcal{F}_8} = (0, 0)$ $\mathbf{p}_8^{\mathcal{F}_8} = (1.2, 0)$ $\mathbf{p}_9^{\mathcal{F}_8} = (0.6, 0.6\sqrt{3})$

sum along a set of independent loops of the mechanism. In such a case, Eq. (2.1) consists of Eq. (3.23) written for all links, together with

$$\mathbf{R}_2\mathbf{p}_4^{\mathcal{F}_2} + \mathbf{R}_5\mathbf{p}_7^{\mathcal{F}_5} + \mathbf{R}_8\mathbf{p}_8^{\mathcal{F}_8} - \mathbf{R}_6\mathbf{p}_8^{\mathcal{F}_6} - \mathbf{R}_3\mathbf{p}_5^{\mathcal{F}_3} = \mathbf{0},$$
$$\mathbf{R}_2\mathbf{p}_4^{\mathcal{F}_2} + \mathbf{R}_5\mathbf{p}_7^{\mathcal{F}_5} + \mathbf{R}_8\mathbf{p}_9^{\mathcal{F}_8} - \mathbf{R}_7\mathbf{p}_9^{\mathcal{F}_7} - \mathbf{R}_4\mathbf{p}_6^{\mathcal{F}_4} = \mathbf{0},$$

which enforce the closure of the two loops in Fig. 3.6. Whereas the original system has 25 equations and 28 variables, this reduced system has only 11 equations and 14 variables, and thus it is easier to solve.

If ω_i is the joint velocity of J_i, and $\hat{\mathbf{S}}_i$ is the associated twist, the twist loop equations relative to loops 1 and 2 can be formulated as

$$\omega_1\hat{\mathbf{S}}_1 + \omega_4\hat{\mathbf{S}}_4 + \omega_7\hat{\mathbf{S}}_7 - \omega_8\hat{\mathbf{S}}_8 - \omega_5\hat{\mathbf{S}}_5 - \omega_2\hat{\mathbf{S}}_2 = \mathbf{0}, \quad (3.24)$$
$$\omega_1\hat{\mathbf{S}}_1 + \omega_4\hat{\mathbf{S}}_4 + \omega_7\hat{\mathbf{S}}_7 - \omega_9\hat{\mathbf{S}}_9 - \omega_6\hat{\mathbf{S}}_6 - \omega_3\hat{\mathbf{S}}_3 = \mathbf{0}, \quad (3.25)$$

and the twist $\hat{\boldsymbol{T}} = (v_x, v_y, \omega)$ of the end-effector can be introduced using the equation

$$\hat{\boldsymbol{T}} = \omega_1\hat{\mathbf{S}}_1 + \omega_4\hat{\mathbf{S}}_4 + \omega_7\hat{\mathbf{S}}_7. \quad (3.26)$$

Hence, the velocity equation $\mathbf{L}\cdot\mathbf{m} = \mathbf{0}$ is in this case defined by Eqs. (3.24)–(3.26), being $\mathbf{m} = (v_x, v_y, \omega, \omega_1, \ldots, \omega_9)$ and

$$\mathbf{L} = \begin{bmatrix} -\mathbf{I}_{3\times3} & \hat{\mathbf{S}}_1 & \mathbf{0} & \mathbf{0} & \hat{\mathbf{S}}_4 & \mathbf{0} & \mathbf{0} & \hat{\mathbf{S}}_7 & \mathbf{0} & \mathbf{0} \\ \mathbf{0} & \hat{\mathbf{S}}_1 & -\hat{\mathbf{S}}_2 & \mathbf{0} & \hat{\mathbf{S}}_4 & -\hat{\mathbf{S}}_5 & \mathbf{0} & \hat{\mathbf{S}}_7 & -\hat{\mathbf{S}}_8 & \mathbf{0} \\ \mathbf{0} & \hat{\mathbf{S}}_1 & \mathbf{0} & -\hat{\mathbf{S}}_3 & \hat{\mathbf{S}}_4 & \mathbf{0} & -\hat{\mathbf{S}}_6 & \hat{\mathbf{S}}_7 & \mathbf{0} & -\hat{\mathbf{S}}_9 \end{bmatrix}. \quad (3.27)$$

Taking into account that $\mathbf{m}_v = (\omega_4, \omega_5, \omega_6)$ and that $\mathbf{m}_u = (v_x, v_y, \omega)$, the matrices $\mathbf{L}_y$ and $\mathbf{L}_z$ needed to formulate Eqs. (2.7) and (2.8) are

$$\mathbf{L}_y = \begin{bmatrix} -\mathbf{I}_{3\times3} & \hat{\mathbf{S}}_1 & \mathbf{0} & \mathbf{0} & \hat{\mathbf{S}}_7 & \mathbf{0} & \mathbf{0} \\ \mathbf{0} & \hat{\mathbf{S}}_1 & -\hat{\mathbf{S}}_2 & \mathbf{0} & \hat{\mathbf{S}}_7 & -\hat{\mathbf{S}}_8 & \mathbf{0} \\ \mathbf{0} & \hat{\mathbf{S}}_1 & \mathbf{0} & -\hat{\mathbf{S}}_3 & \hat{\mathbf{S}}_7 & \mathbf{0} & -\hat{\mathbf{S}}_9 \end{bmatrix}, \quad (3.28)$$

and

$$\mathbf{L}_z = \begin{bmatrix} \hat{\mathbf{S}}_1 & \mathbf{0} & \mathbf{0} & \hat{\mathbf{S}}_4 & \mathbf{0} & \mathbf{0} & \hat{\mathbf{S}}_7 & \mathbf{0} & \mathbf{0} \\ \hat{\mathbf{S}}_1 & -\hat{\mathbf{S}}_2 & \mathbf{0} & \hat{\mathbf{S}}_4 & -\hat{\mathbf{S}}_5 & \mathbf{0} & \hat{\mathbf{S}}_7 & -\hat{\mathbf{S}}_8 & \mathbf{0} \\ \hat{\mathbf{S}}_1 & \mathbf{0} & -\hat{\mathbf{S}}_3 & \hat{\mathbf{S}}_4 & \mathbf{0} & -\hat{\mathbf{S}}_6 & \hat{\mathbf{S}}_7 & \mathbf{0} & -\hat{\mathbf{S}}_9 \end{bmatrix}. \quad (3.29)$$

Finally, it is easy to see that in planar mechanisms the expression of $\hat{\mathbf{S}}_i$ in Table 3.1 reduces to

$$\hat{\mathbf{S}}_i = \begin{bmatrix} y_i \\ -x_i \\ 1 \end{bmatrix}, \tag{3.30}$$

where $\mathbf{p}_i = (x_i, y_i)$ are the absolute coordinates of P_i [17]. These coordinates can be expressed in terms of the configuration of the mechanism using Eq. (3.17), if the $\mathbf{r}_j$ variables have not been eliminated, or by resorting to the expressions

$$\begin{aligned} \mathbf{p}_4 &= \mathbf{p}_1 + \mathbf{R}_2 \mathbf{p}_4^{\mathcal{F}_2}, & \mathbf{p}_7 &= \mathbf{p}_4 + \mathbf{R}_5 \mathbf{p}_7^{\mathcal{F}_5}, \\ \mathbf{p}_5 &= \mathbf{p}_2 + \mathbf{R}_3 \mathbf{p}_5^{\mathcal{F}_3}, & \mathbf{p}_8 &= \mathbf{p}_5 + \mathbf{R}_6 \mathbf{p}_8^{\mathcal{F}_6}, \\ \mathbf{p}_6 &= \mathbf{p}_3 + \mathbf{R}_4 \mathbf{p}_6^{\mathcal{F}_4}, & \mathbf{p}_9 &= \mathbf{p}_6 + \mathbf{R}_7 \mathbf{p}_9^{\mathcal{F}_7}, \end{aligned}$$

otherwise. Note that $\mathbf{p}_1 = \mathbf{p}_1^{\mathcal{F}_1}$, $\mathbf{p}_2 = \mathbf{p}_2^{\mathcal{F}_1}$, and $\mathbf{p}_3 = \mathbf{p}_3^{\mathcal{F}_1}$, so these vectors take the constant values given in Table 3.2.

We now know how to formulate Eqs. (2.7) and (2.8) in the form required by the numerical method. In order to generate the portraits mentioned in Sect. 3.4, however, it is convenient to introduce the input and output variables in the systems. The output variables are given by $\mathbf{u} = (x_7, y_7, \theta_8)$, and they are already explicit in the formulation. The input variables are the relative angles $\mathbf{v} = (\alpha_4, \alpha_5, \alpha_6)$ indicated in Fig. 3.6, whose sines and cosines can be obtained by completing the system with the equations

$$\begin{aligned} s_{\alpha_j} &= s_{j+1} c_{j-2} - c_{j+1} s_{j-2}, \\ c_{\alpha_j} &= c_{j+1} c_{j-2} + s_{j+1} s_{j-2}, \end{aligned}$$

for $j = 4, 5, 6$, which are the algebraic version of $\alpha_j = \theta_{j+1} - \theta_{j-2}$. By introducing these equations, the numerical method will be able to provide ranges for the sines and cosines of α_4, α_5, and α_6, from which it is straightforward to obtain the corresponding angular ranges.

The singularity surfaces obtained by the method are shown in Fig. 3.8 (top) projected to the output space. The blue surface corresponds to the inverse singularity locus, which provides the boundaries of the workspace. The red surface corresponds to the forward singularity locus, i.e. to configurations where the motion control is compromised, due to the specific choice of actuated degrees-of-freedom. Even though these singularity surfaces appear to be complex, it can be shown that the constant-orientation slices of the forward singularity locus are conic sections in the (x_7, y_7) plane [22, 30]. Any of these slices can be readily obtained using the proposed method by simply fixing the value of θ_8 in the equations, obtaining the red curves shown in Fig. 3.8 (bottom), where only pairs of lines, parabolas, or ellipses appear as expected. The inverse singularity curves in such plots also coincide with those obtained through the intersection of vertex spaces [22, 31].

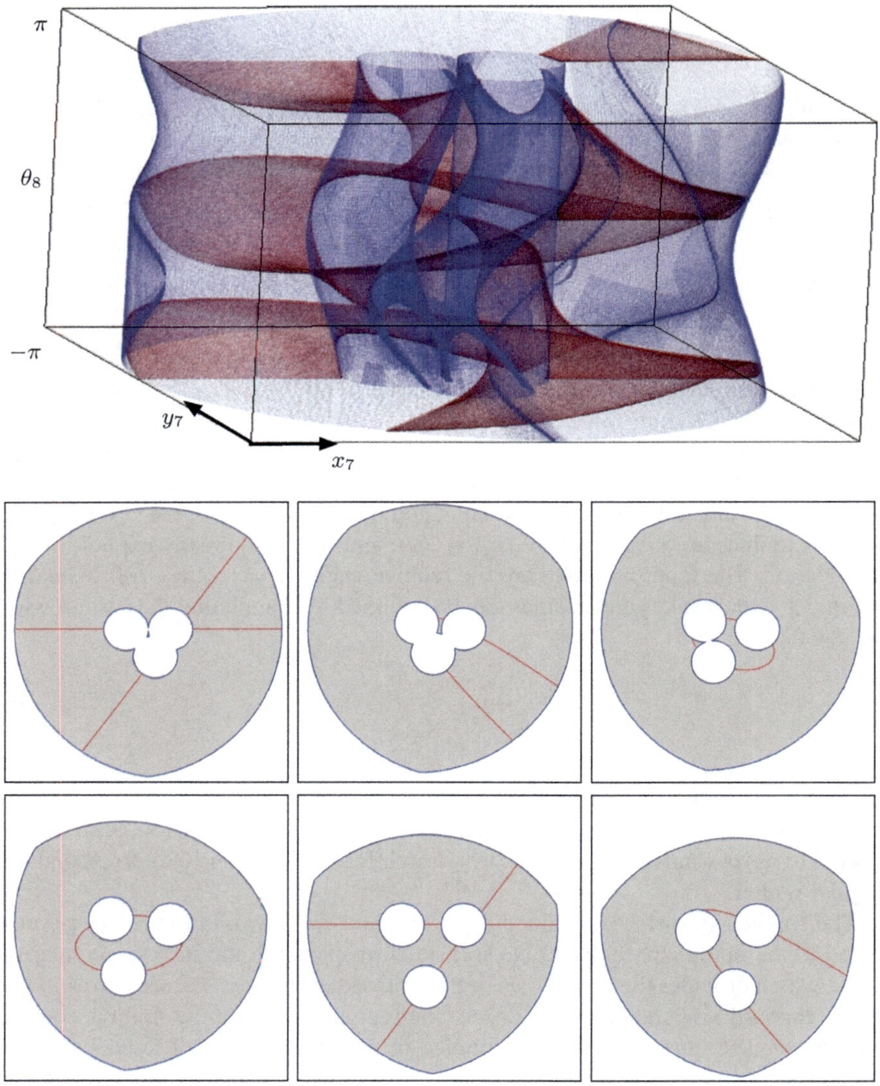

Fig. 3.8 Output portrait obtained for the 3-R$\underline{\text{R}}$R mechanism. (*Top*) Forward (*red*) and inverse (*blue*) singularity surfaces in the (x_7, y_7, θ_8) space. The boxes are drawn with translucent faces to better appreciate the shape of the surfaces. See [33] for an animated version of this figure. (*Bottom*) Slices of the output portrait at a constant value of θ_8. From *left* to *right*, and *top* to *bottom*, the values of θ_8 are $-\pi$, $-\frac{3\pi}{4}$, $-\frac{\pi}{2}$, $-\frac{\pi}{4}$, 0, and $\frac{\pi}{4}$

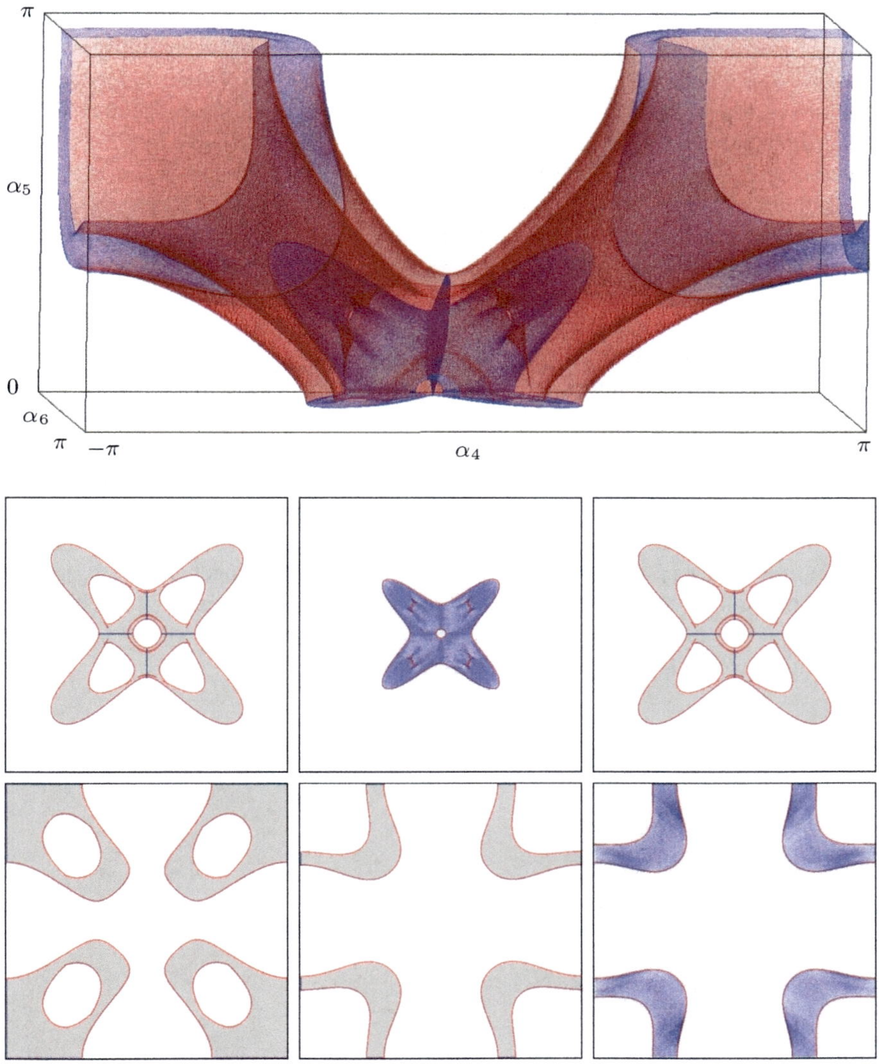

Fig. 3.9 Input portrait of the 3-R$\underline{\text{R}}$R mechanism. (*Top*) Forward (*red*) and inverse (*blue*) singularity surfaces in the $(\alpha_4, \alpha_5, \alpha_6)$ space. Only two octants of the space are shown for simplicity, since the other octants can be obtained by symmetry. See [33] for an animated version of this figure. (*Bottom*) Slices of the input portrait at different values of α_5. From *left* to *right*, and *top* to *bottom*, the values of α_5 are $-\frac{\pi}{4}$, 0, $\frac{\pi}{4}$, $\frac{\pi}{2}$, $\frac{3\pi}{4}$, and π

By simply changing the projection coordinates we can easily represent $\mathcal{S}$ in the input space as well, obtaining the results shown in Fig. 3.9. The forward singularities here delimit the motion range of the actuators, and it can be seen how the inverse singularities only appear in planes where one of the α_i angles is either 0 or π, in agreement with the fact that the platform only loses instantaneous mobility when at least one of the legs is fully extended or folded back [22]. To better understand the structure of the singularity surface on the input space, some slices are also shown for constant values of α_6. Observe how the whole region attainable by the inputs is singular for $\alpha_6 = 0$ or $\alpha_6 = \pi$. In these slices the inverse singularities are no longer one-dimensional as one would expect. Whereas this circumstance poses no problem to the proposed method, it may indeed hinder the application of other methods relying on a discretisation of α_6.

It must be noted that the structure of the singularity set can become quite complex even in simple mechanisms. For example, if in the 3-RRR mechanism we mount the actuators in the ground joints J_1, J_2, and J_3 instead of in the intermediate ones, we obtain the 3-$\underline{\text{R}}$RR architecture, and the constant-orientation slices of the forward singularity locus are then described by polynomials in x_7 and y_7 of minimal degree 42 [22]. Polynomials of such kind constitute valuable tools for the analysis of the singularity set, but their derivation often requires quite involved manipulations guided by intuition [23, 24, 27, 34], which complicates the application of such strategy to every new mechanism that has to be analysed. The proposed method can compute the mentioned slices just as easily as in the case of the 3-R$\underline{\text{R}}$R mechanism (Fig. 3.10). The full potential of the method is even more apparent in mechanisms of much higher complexity, like the one in Sect. 4.5.4, in which the approach based on descriptive polynomials would be rather difficult to apply.

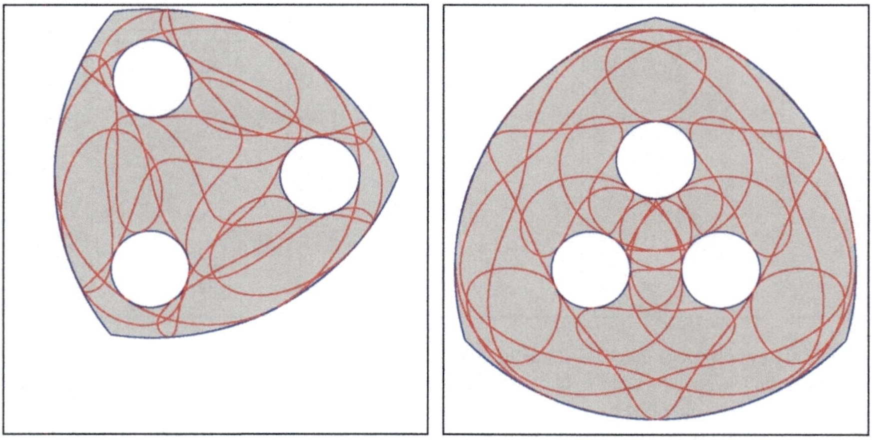

Fig. 3.10 Slices of the output portrait of the 3-$\underline{\text{R}}$RR mechanism computed for fixed values of the platform angle θ_8. We have used the parameters in Table 3.2, and the values $\theta_8 = -\frac{\pi}{4}$ (*left plot*) and $\theta_8 = 0$ (*right plot*). The plot for $\theta_8 = 0$ agrees with the one published in [22, 32]

3.5.2 The Gough-Stewart Platform

The Gough-Stewart platform consists of a moving plate connected to a fixed base by means of six legs, where each leg is a spherical-prismatic-universal chain. The most frequent design follows an octahedral pattern (Fig. 3.11), but the leg anchor points may all be different in general, and not necessarily coplanar [35]. The six prismatic joints are actuated, which allows one to control the six degrees of freedom of the platform independently, and the remaining joints are passive. Since the pioneering work by Gough, Cappel, and Stewart (see [36] for an interesting review of the history of the invention), this hexapodal architecture has been used in many applications, ranging from high-performance micro- and nanopositioning systems, to huge flight and driving simulators (Fig. 3.12).

The assembly constraints can be formulated as explained in Sect. 3.2. We define two links, the ground base and the moving platform, labelled L_1 and L_2, and the centre points of the spherical and universal joints, $P_{1,i}$ and $P_{2,i}$, for each leg from $i = 1$ to 6. We also let O be the origin of the absolute frame, $\mathcal{F}_1$, and P be that of the frame attached to the platform, $\mathcal{F}_2$ (Fig. 3.11). By taking into account the constraints imposed by the universal, prismatic, and spherical joints of the i-th leg, i.e. Eqs. (3.3)–(3.7), the constraint imposed on the platform by such leg can be reduced to

$$\mathbf{r}_P = \mathbf{p}_{1,i} + d_i \mathbf{d}_i - \mathbf{R}_2 \mathbf{p}_{2,i}^{\mathcal{F}_2}, \tag{3.31}$$

$$\|\mathbf{d}_i\|^2 = 1, \tag{3.32}$$

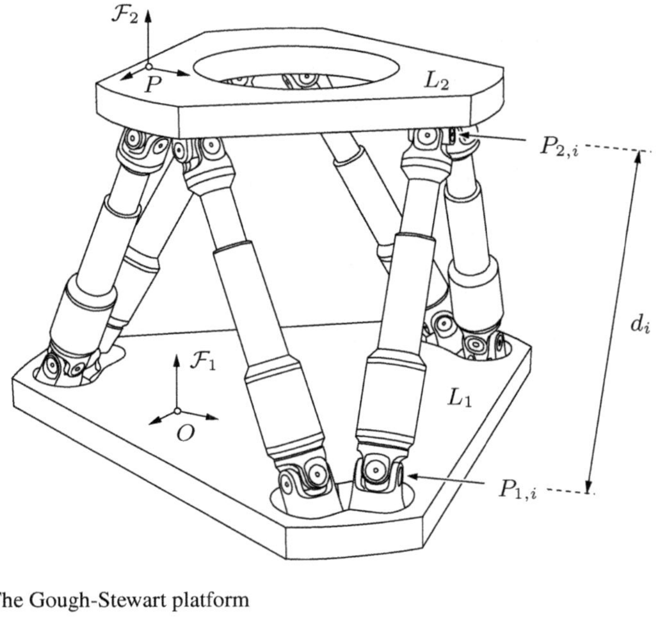

Fig. 3.11 The Gough-Stewart platform

Fig. 3.12 Driving Simulator at Toyota's Higashi-Fuji Technical Center in Susono, Shizuoka, Japan. The 7.1 m dome sits on a hydraulic Gough-Stewart Platform installed on top of a large excursion X-Y system. Inside the dome, a Lexus LS body is mounted on a turntable and a 360° projection system produces a high-resolution driving environment. (Courtesy of Toyota Motor Corporation.)

where $\mathbf{r}_P$, $\mathbf{p}_{1,i}$, and $\mathbf{p}_{2,i}$ are the position vectors of points P, $P_{1,i}$, and $P_{2,i}$, respectively, and $\mathbf{d}_i$ is a unit vector along the i-th leg. Also, d_i is the length of the leg, representing the displacement of the prismatic joint. Thus, Eq. (2.1) can be written by collecting Eq. (3.31) and (3.32) for all legs, and Eqs (3.8)–(3.11) for $\mathbf{R}_2$, obtaining a system of 30 equations in 36 variables where the pose of the end-effector is represented by $(\mathbf{r}_P, \mathbf{R}_2)$. The geometric parameters used here are given in Table 3.3 and correspond to the mechanism studied in [23].

It can be seen that each pair of legs forms a closed kinematic chain of the mechanism, but that only five of them are independent. Therefore, the velocity equation can be formulated by writing Eq. (3.12) for five distinct leg pairs, and Eq. (3.13) for any leg. However, in parallel mechanisms like the Gough-Stewart platform, the velocity equation can also be obtained by writing Eq. (3.13) for each leg [15]. In such a case, the matrix $\mathbf{L}$ adopts the form

$$\mathbf{L} = \begin{bmatrix} -\mathbf{I}_{6\times6} & \hat{\mathbf{S}}_1^a & 0 & 0 & 0 & 0 & 0 & \hat{\mathbf{S}}_1^p & 0 & 0 & 0 & 0 & 0 \\ -\mathbf{I}_{6\times6} & 0 & \hat{\mathbf{S}}_2^a & 0 & 0 & 0 & 0 & 0 & \hat{\mathbf{S}}_2^p & 0 & 0 & 0 & 0 \\ -\mathbf{I}_{6\times6} & 0 & 0 & \hat{\mathbf{S}}_3^a & 0 & 0 & 0 & 0 & 0 & \hat{\mathbf{S}}_3^p & 0 & 0 & 0 \\ -\mathbf{I}_{6\times6} & 0 & 0 & 0 & \hat{\mathbf{S}}_4^a & 0 & 0 & 0 & 0 & 0 & \hat{\mathbf{S}}_4^p & 0 & 0 \\ -\mathbf{I}_{6\times6} & 0 & 0 & 0 & 0 & \hat{\mathbf{S}}_5^a & 0 & 0 & 0 & 0 & 0 & \hat{\mathbf{S}}_5^p & 0 \\ -\mathbf{I}_{6\times6} & 0 & 0 & 0 & 0 & 0 & \hat{\mathbf{S}}_6^a & 0 & 0 & 0 & 0 & 0 & \hat{\mathbf{S}}_6^p \end{bmatrix}, \quad (3.33)$$

where $\hat{\mathbf{S}}_i^a$ is the unit twist of the prismatic joint of the i-th leg, and $\hat{\mathbf{S}}_i^p$ is a 6×5 matrix whose columns are the unit twists of the universal and spherical joints of this leg. Then, the velocity vector $\mathbf{m}$ contains the six components of the output twist, the six active velocities of the prismatic joints, and the 30 passive joint velocities of the universal and spherical joints. Altogether, this is a 36×42 matrix, but a method explained in [15] allows us to reduce its size by eliminating the passive joint velocities of the equation. The method consists in multiplying each side of Eq. (3.13) of each leg by a unit screw reciprocal to all passive joints of the leg. Note that, in general, such a screw is any zero-pitch screw that passes through the centres of the universal

Table 3.3 Parameters of the analysed Gough-Stewart platform, in millimeters

Leg	Base (L_1)	Platform (L_2)
1	$\mathbf{p}_{1,1}^{\mathcal{F}1} = (92.58, 99.64, 20.10)$	$\mathbf{p}_{2,1}^{\mathcal{F}2} = (30.00, 73.00, -35.10)$
2	$\mathbf{p}_{1,2}^{\mathcal{F}1} = (132.58, 30.36, 28.45)$	$\mathbf{p}_{2,2}^{\mathcal{F}2} = (78.22, -10.52, -23.00)$
3	$\mathbf{p}_{1,3}^{\mathcal{F}1} = (40.00, -120.00, 31.18)$	$\mathbf{p}_{2,3}^{\mathcal{F}2} = (48.22, -62.48, -33.60)$
4	$\mathbf{p}_{1,4}^{\mathcal{F}1} = (-46.00, -130.00, 3.10)$	$\mathbf{p}_{2,4}^{\mathcal{F}2} = (-44.22, -56.48, -25.50)$
5	$\mathbf{p}_{1,5}^{\mathcal{F}1} = (-130.00, 23.36, 13.48)$	$\mathbf{p}_{2,5}^{\mathcal{F}2} = (-70.22, -20.52, -34.10)$
6	$\mathbf{p}_{1,6}^{\mathcal{F}1} = (-82.58, 89.77, 8.76)$	$\mathbf{p}_{2,6}^{\mathcal{F}2} = (-34.00, 45.00, -39.00)$

and spherical joints [37, Sect. 5.5.2], for instance[1]

$$\hat{s}_i^r = \begin{bmatrix} \mathbf{d}_i \\ \mathbf{d}_i \times \mathbf{p}_{1,i} \end{bmatrix}. \tag{3.34}$$

By doing the multiplication, Eq. (3.13) reduces to

$$\hat{s}_i^{r\mathsf{T}} \hat{\boldsymbol{T}} = \omega_i^a, \tag{3.35}$$

where ω_i^a is the velocity of the prismatic joint of leg i, because $\hat{s}_i^{r\mathsf{T}} \hat{\mathbf{S}}_i^p = \mathbf{0}$ and $\hat{s}_i^{r\mathsf{T}} \hat{\mathbf{S}}_i^a = 1$. Collecting Eq. (3.35) for all legs we obtain the velocity equation as

$$\mathbf{J}^\mathsf{T} \hat{\boldsymbol{T}} = \mathbf{m}_\upsilon, \tag{3.36}$$

where $\mathbf{J}$ is a matrix that has the unit screws $\hat{s}_i^r$ as columns, and $\mathbf{m}_\upsilon$ contains the six ω_i^a.

For some configurations, the space of reciprocal screws of a given leg may have dimension larger than one. In such a case, the corresponding Eq. (3.13) of the leg should be multiplied by all vectors of a basis of the space of reciprocal screws, resulting in a velocity equation of the form

$$\mathbf{J}^\mathsf{T} \hat{\boldsymbol{T}} = \mathbf{H}\mathbf{m}_\upsilon, \tag{3.37}$$

where $\mathbf{J}^\mathsf{T}$ and $\mathbf{H}$ are rectangular matrices with more rows than columns. For the Gough-Stewart platform this can happen only when the spaces of rotations of the universal and spherical joints have a nonzero intersection, i.e. when the centre of the spherical joint is in the plane of the two universal joint axes, and the leg is said to be singular. This situation corresponds to an RPM type singularity, but it rarely occurs in practice because such configurations are commonly unfeasible due to joint limits. In the usual case, $\mathbf{J}$ remains square, $\mathbf{H}$ is the identity matrix because of the choice of $\hat{s}_i^r$, and Eq. (3.37) reduces to Eq. (3.36), which is the classical simplified version of the velocity equation [38].

Based on Eq. (3.36) we see that the solution to the FIKP involves the inverse of $\mathbf{J}$, so that the forward singularities can be characterised by those configurations where $\mathbf{J}$ is not full rank. Thus, in order to compute the forward singularities, $\mathbf{J}$ plays the role of $\mathbf{L}_y$ in Eq. (2.7), obtaining a more compact system of equations than using $\mathbf{L}_y$ directly from Eq. (3.33). A similar reasoning applies for the inverse singularities, which can be computed as the solution of Eq. (2.8) replacing $\mathbf{L}_z$ by $\mathbf{H}$. We shall here assume here that the legs are never singular by design, and hence we shall only compute the forward singularity locus.

[1] The lowercase notation $\hat{s}_i$ indicates that the screw is written in ray coordinates [17], i.e., in "(vector, moment)" form, to allow the reciprocal product.

Being five-dimensional, the forward singularity locus cannot be directly visualised in this case. However, insight can be gained by computing its three-dimensional slices. In order to plot these slices when they involve the orientation of the platform we introduce three-parameter expressions for $\mathbf{R}_2$ in the system of equations to be solved. We can use any possible parameterisation, including those based on conventional Euler angles, tilt-and-torsion angles, or Euler-Rodrigues parameters [37, 39]. For ease of comparison with [23] we here adopt the parameterisation provided by roll (ϕ), pitch (θ), and yaw (ψ) angles, for which

$$\mathbf{R}_2 = \mathbf{R}_z(\psi)\mathbf{R}_y(\theta)\mathbf{R}_x(\phi)$$

i.e.,

$$\mathbf{R}_2 = \begin{bmatrix} c_\theta c_\psi & s_\phi s_\theta c_\psi - c_\phi s_\psi & c_\phi s_\theta c_\psi + s_\phi s_\psi \\ c_\theta s_\psi & s_\phi s_\theta s_\psi + c_\phi c_\psi & c_\phi s_\theta s_\psi - s_\phi c_\psi \\ -s_\theta & s_\phi c_\theta & c_\phi c_\theta \end{bmatrix},$$

where $c_\alpha = \cos\alpha$ and $s_\alpha = \sin\alpha$. The addition of such parameterisations introduces trilinear terms in the equations, but these can always be made bilinear by using proper changes of variables.

Two slices of the singularity locus are shown in Fig. 3.13, one computed at a constant orientation of the platform, with $\phi = -2°$, $\theta = 30°$, $\psi = -87°$, and one at a constant position of point P in the platform, whose position vector is set to $\mathbf{r}_P = (10, 10, 10)$. Note that any other slice obtained by keeping constant three or more platform pose variables could also be computed just as easily.

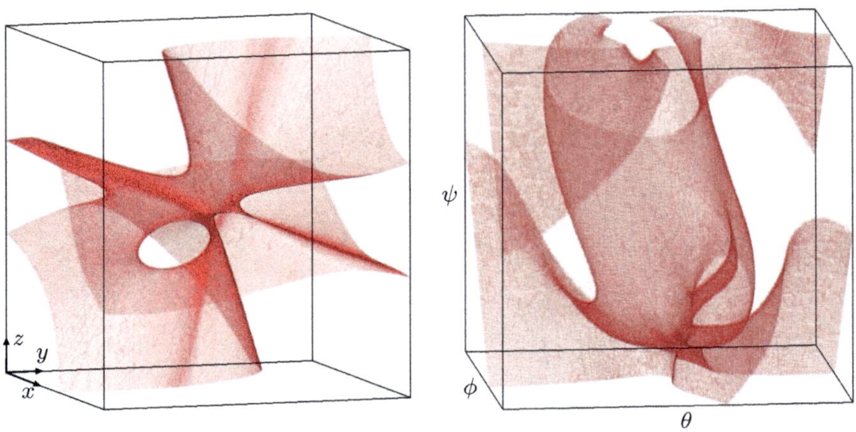

Fig. 3.13 Forward singularities of the Gough-Stewart platform with its orientation fixed using $\phi = -2°$, $\theta = 30°$, and $\psi = -87°$ (*left*), and with point P fixed at $\mathbf{r}_P = (10, 10, 10)$ (*right*). The position variables and the orientation angles have been limited to the ranges $[-100, 100]$ and $[-90°, 90°]$. These surfaces can be seen under rotation in [33]

3.5.3 A Double-Loop Mechanism

The 2-DOF mechanism shown in Fig. 3.14 is used here to illustrate the computation of each one of the six lower-level singularity sets. The inputs of the mechanism are the joint velocities at A and E, and the output is the velocity of point G. By gathering the loop-closure equations of the mechanism, and introducing two further equations to include the position of G, Eq. (2.1) can be formulated as follows:

$$\left.\begin{aligned} \cos\theta_A + \cos\theta_B - 2\cos\theta_D - 1 &= 0 \\ \sin\theta_A + \sin\theta_B - 2\sin\theta_D &= 0 \\ 2\cos\theta_D + \tfrac{3}{2}\cos\theta_C + 2\cos\theta_G - 3\cos\theta_E - 1 &= 0 \\ 2\sin\theta_D + \tfrac{3}{2}\sin\theta_C + 2\sin\theta_G - 3\sin\theta_E &= 0 \\ -x + 2\cos\theta_D + \tfrac{3}{2}\cos\theta_C &= 0 \\ -y + 2\sin\theta_D + \tfrac{3}{2}\sin\theta_C &= 0 \end{aligned}\right\}, \tag{3.38}$$

where θ_A, θ_B, θ_C, θ_D, θ_E and θ_G are the counterclockwise angles of links AB, BC, CG, DC, EF, and GF, respectively, relative to the ground, and x and y are the coordinates of point G relative to a fixed frame centred in D. In this case, the velocity equation of the mechanism is easily obtained by differentiating Eq. (3.38) with respect to all variables, although the twist loop equations could also be used. In any case, the systems of equations encoding the six singularity types, Eq. (2.12)–(2.17), can be written and expanded to the desired quadratic form of Eq. (3.20) by introducing the changes of variables

$$\begin{aligned} c_\tau &= \cos\theta_\tau \\ s_\tau &= \sin\theta_\tau \end{aligned}$$

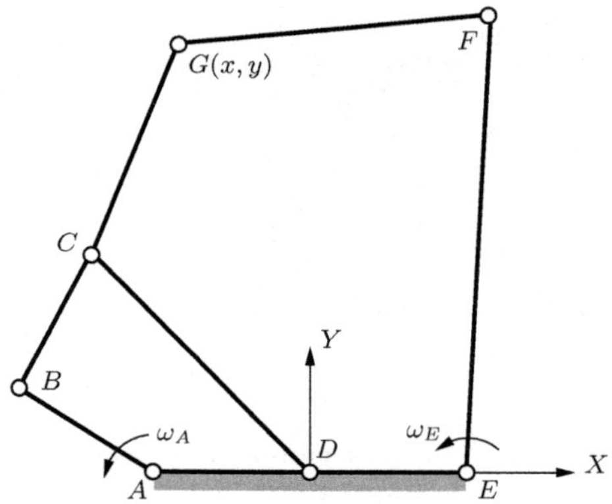

Fig. 3.14 A 2-DOF planar mechanism. The link dimensions are $AB = AD = BC = DE = 1$, $CD = FG = 2$, $CG = 1.5$ and $EF = 3$

together with the equation

$$c_\tau^2 + s_\tau^2 = 1,$$

for all angles with $\tau = \{A, B, C, D, E, G\}$.

Given that the mechanism has two degrees of freedom, its C-space is a surface, which is shown projected to the x, y, and θ_A variables in Fig. 3.15. This surface is obtained from the computation of all solutions of Eq. (2.1) using the same numerical technique presented in Sect. 3.3. Note that by fixing x, y, and θ_A, there are still two possible positions of point F, so that most of the points in this projection correspond, in fact, to two different configurations of the mechanism. Only the points where E, F, and G are aligned represent a single configuration, and these are exactly the boundaries of the two "holes" that the surface presents.

The singularity set is generally of lower dimension than the C-space, so that only curves or points are to be expected in the solution set of all systems of equations. The result of the computation of each singularity type defined in Sect. 2.3 is shown in Figs. 3.16 and 3.17, projected to the output and one input (x, y, θ_A), and onto the output only, respectively. In Fig. 3.16, the C-space is shown in blue, separated in two parts so that a cross-section can be seen, but both parts are actually connected through π and $-\pi$ as in Fig. 3.15. The grey area in Fig. 3.17 represents all attainable positions of point G, i.e. the workspace of the mechanism.

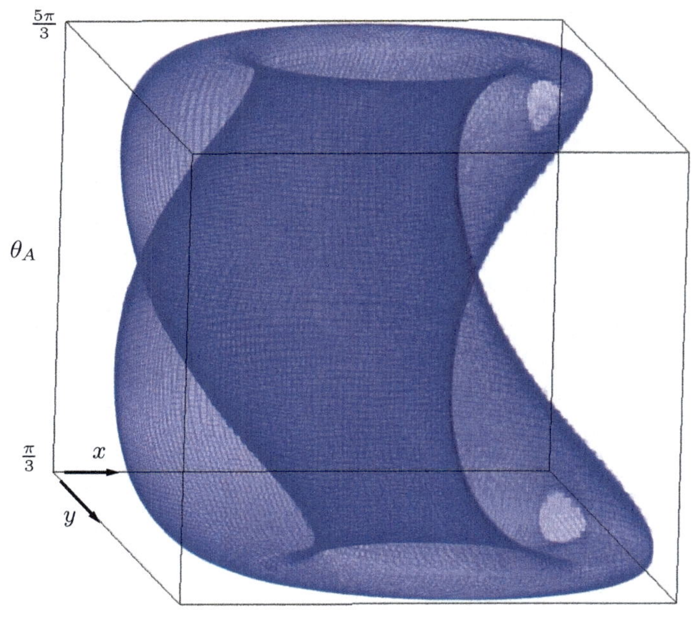

Fig. 3.15 Two-dimensional C-space of the mechanism in Fig. 3.14 computed with $\sigma = 0.1$. Two holes can be seen, whose boundary corresponds to configurations where E, F, and G are aligned

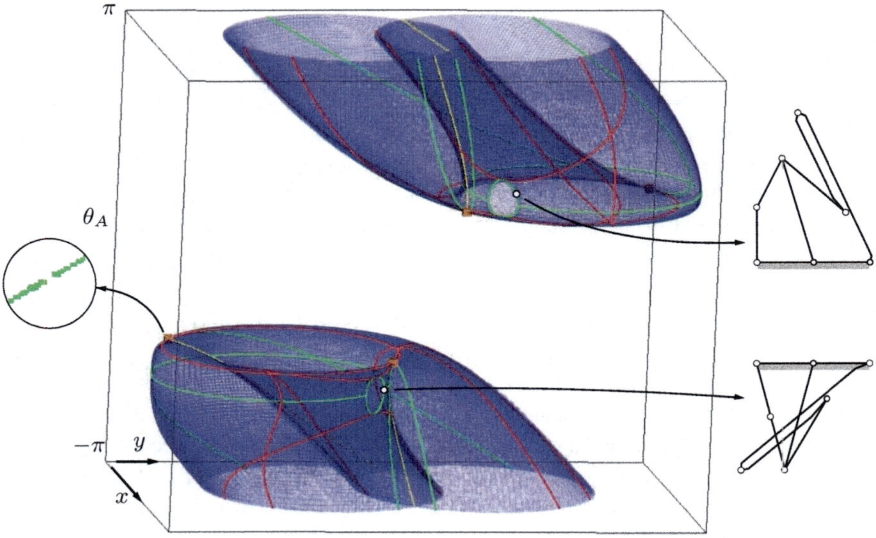

Fig. 3.16 The singular configurations of the mechanism in Fig. 3.14 shown overlaid on a projection of its C-space to the x, y and θ_A variables. Different *colors* are used to identify the several singularity types encountered: *green* for the RI and IO types, *red* for the RO and II types, and *orange* for the RPM type

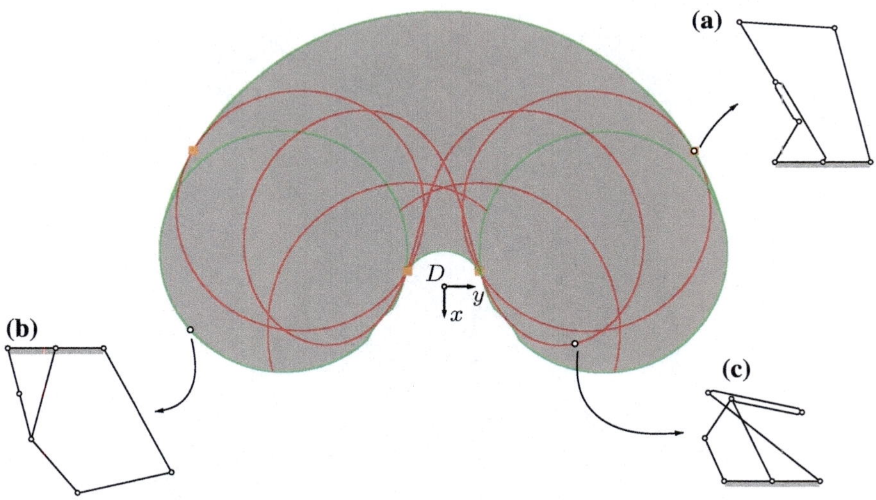

Fig. 3.17 A projection of the plot in Fig. 3.16 to the (x, y) plane. (**a**) A singularity of RPM, IO, and II type. (**b**) A singularity of RI and IO type. (**c**) A singularity of RO and II type

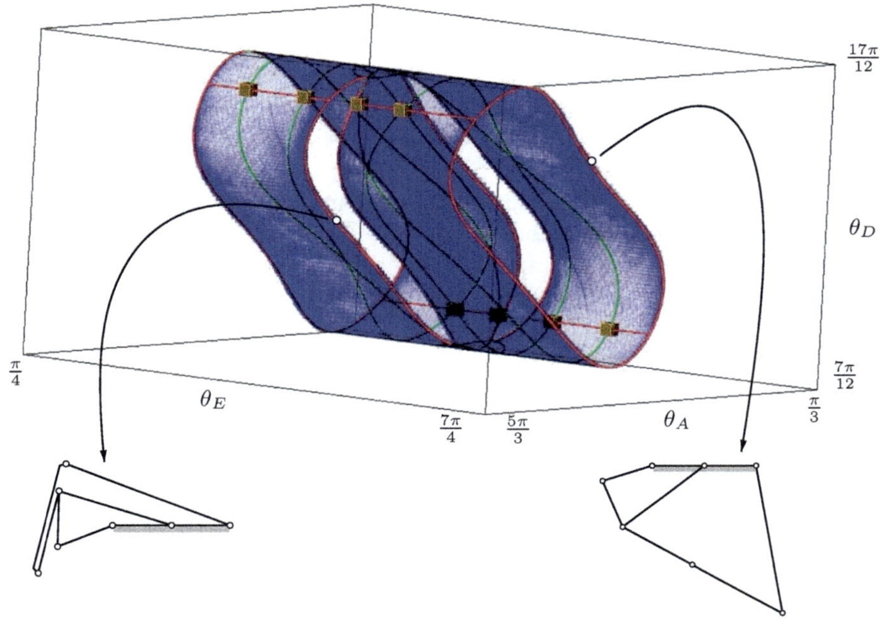

Fig. 3.18 A projection of the C-space and the computed singularities to the $(\theta_A, \theta_E, \theta_D)$ space, together with two configurations where points C, G, and F are aligned. *Green* corresponds to the RI and IO types, *red* to the RO and II types, and *orange* to the RPM type. There are no singularities of IIM type

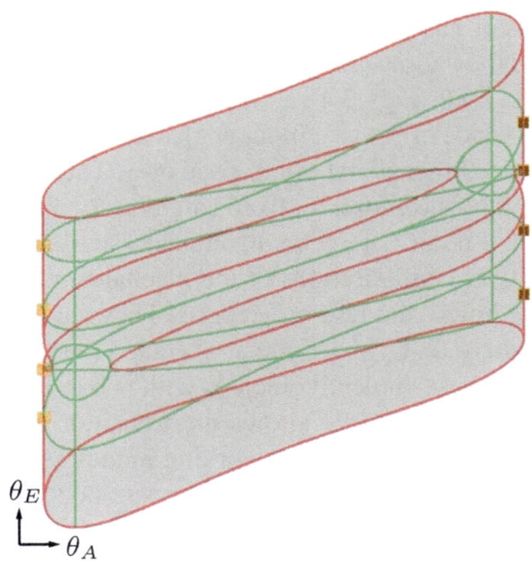

Fig. 3.19 A projection of the plot in Fig. 3.18 to the (θ_A, θ_E) space

As it turns out, this mechanism contains no IIM type configurations, and the computation of this type of singularity gives no box as output. On the contrary, there are eight distinct RPM type singularities, which in these projections appear coincident in pairs as four orange boxes, corresponding to the two possible locations of F. Using a different projection, for instance to $(\theta_A, \theta_D, \theta_E)$, the eight boxes appear separated (Fig. 3.18).

The green curves correspond to singularities that are both of RI and IO type. These configurations can be seen to contour the two "holes" of the C-space in the projection of Fig. 3.16. The red curves correspond to configurations simultaneously belonging to the RO and II type. Even though the curves for RI and IO seem to coincide everywhere, there are some IO configurations that are not of RI type, and the same happens for II and RO singularities, respectively. This is illustrated in Fig. 3.16 with a magnifying bubble that shows only the output of computing RI type singularities. These gaps in the curves of RI and RO, which can be found by properly adjusting the ϵ parameter, coincide with the location of the RPM singularities and, hence, the RPM singularities are also of II and IO type but not of RI or RO type. Figure 3.17a shows an example of an (RPM, II, IO) singularity, while Fig. 3.17b, c show examples of (RI, IO) and (RO, II) singularities, respectively.

Yellow curves corresponding to configurations where points D, B, and G are aligned are also shown in Fig. 3.16. For x, y and θ_A fixed, there is only one possible location of point C in general, but in these configurations there are two possibilities. An exception exists, however, when C is also aligned with D, B, and G, which corresponds to the singular configurations of RPM type. Hence, these configurations, together with those where E, F, and G are aligned, allow the transition between different working modes. In fact, the configurations of the yellow curves coincide with the points of self-intersection of the projection of the C-space to the (x, y, θ_A) space. The C-space itself has no self-intersections as there are no C-space, or IIM type, singularities, and the yellow points are only singularities of the projection map.

Using the same color code, Figs. 3.18 and 3.19 show the projection of the results to the three-dimensional space of the two input angles and one passive joint angle (θ_A, θ_E, θ_D) and onto the two-dimensional input space only. The eight RPM singularities appear separated. Similarly as before, for fixed values of θ_A, θ_E, and θ_D, there are still two possible locations of point G in general, and almost all points in this projection correspond to two distinct configurations of the mechanism. It can be seen that the C-space presents four "holes" in these projections. These four contours are made of those configurations where G, C, and F are aligned and there is only one possibility for G. Note that none of these "holes" coincides with one in the previous projection, but, once again, crossing each curve allows the transition between two different working modes. One can imagine the two working modes as the two "sides" of the surface of the C-space projection. To "get to the opposite side", i.e. to change working mode, the motion curve must "go through a hole".

Table 3.4 summarises the performance data of the numerical methods relative to the reported examples.

Table 3.4 Performance data relative to the reported examples

Figures	Mechanism	Set	Slice	d	N_{eq}-N_{var}	σ	T	N_{box}
3.8 3.9	3-R$\underline{\text{R}}$R	Forward singularities	–	2	29–31	0.10	817	148473
			$\theta_8 = -\pi$	1	28–29	0.01	18	2692
			$\theta_8 = -\frac{3\pi}{4}$	1	28–29	0.01	14	1372
			$\theta_8 = -\frac{\pi}{2}$	1	28–29	0.01	12	894
			$\theta_8 = -\frac{\pi}{4}$	1	28–29	0.01	13	1113
			$\theta_8 = 0$	1	28–29	0.01	17	2612
			$\theta_8 = \frac{\pi}{4}$	1	28–29	0.01	14	1658
			$\alpha_6 = -\frac{\pi}{4}$	1	28–29	0.01	186	22195
			$\alpha_6 = 0$	1	28–29	0.01	216	10158
			$\alpha_6 = \frac{\pi}{4}$	1	28–29	0.01	198	22151
			$\alpha_6 = \frac{\pi}{2}$	1	28–29	0.01	118	23654
			$\alpha_6 = \frac{3\pi}{4}$	1	28–29	0.01	55	13578
			$\alpha_6 = \pi$	1	28–29	0.01	53	11950
		Inverse singularities	–	2	29–31	0.10	316	182560
			$\theta_8 = -\pi$	1	28–29	0.01	65	9652
			$\theta_8 = -\frac{3\pi}{4}$	1	28–29	0.01	61	8828
			$\theta_8 = -\frac{\pi}{2}$	1	28–29	0.01	63	8725
			$\theta_8 = -\frac{\pi}{4}$	1	28–29	0.01	51	7748
			$\theta_8 = 0$	1	28–29	0.01	49	7419
			$\theta_8 = \frac{\pi}{4}$	1	28–29	0.01	46	7579
			$\alpha_6 = -\frac{\pi}{4}$	1	28–29	0.01	15	6655
			$\alpha_6 = 0$	2	28–29	0.10	489	106792
			$\alpha_6 = \frac{\pi}{4}$	1	28–29	0.01	15	6653
			$\alpha_6 = \frac{\pi}{2}$	1	28–29	0.01	18	9851
			$\alpha_6 = \frac{3\pi}{4}$	1	28–29	0.01	12	5885
			$\alpha_6 = \pi$	2	28–29	0.10	447	170170
3.10	3-$\underline{\text{R}}$RR	Forward singularities	$\theta = -\frac{\pi}{4}$	1	22-23	0.01	9	9276
			$\theta_8 = 0$	1	22-23	0.01	15	14548
			$\theta_8 = \frac{\pi}{4}$	1	22-23	0.01	10	9335
		Inverse singularities	$\theta = -\frac{\pi}{4}$	1	22-23	0.01	59	19906
			$\theta_8 = 0$	1	22-23	0.01	66	18917
			$\theta_8 = \frac{\pi}{4}$	1	22-23	0.01	51	19998
3.13	Stewart platform	Forward singularities	Fixed orient.	2	25–27	0.02	79	146420
			Fixed position	2	37–39	0.25	2554	195982

(continued)

Table 3.4 (continued)

Figures	Mechanism	Set	Slice	d	N_{eq}-N_{var}	σ	T	N_{box}
3.15 to 3.19	Double-loop mechanism	C-space	–	2	12–14	0.10	31	85238
		RI	–	1	19–20	0.01	12	14903
		RO	–	1	19–20	0.01	12	12773
		IO	–	1	19–20	0.01	14	14906
		II	–	1	19–20	0.01	13	13062
		RPM	–	0	19–18	0.01	4	8
		IIM	–	0	21–20	0.01	2	0

For each set we indicate whether the data refers to the computation of a slice or the whole set, the dimension of the set (d), the number of equations (N_{eq}) and variables (N_{var}) in its defining system, the accuracy threshold assumed (σ), the time employed in computing the set in seconds (T), and the number of solution boxes obtained (N_{box})

References

1. B. Sturmfels, *Solving Systems of Polynomial Equations*, vol. 97 (American Mathematical Society, 2002)
2. D. Cox, J. Little, D. O'Shea, *Ideals, Varieties, and Algorithms: An Introduction to Computational Algebraic Geometry and Commutative Algebra*, vol. 10 (Springer, 2007)
3. A. Morgan, *Solving Polynominal Systems Using Continuation for Engineering and Scientific Problems*, vol. 57 (Society for Industrial Mathematics, 2009)
4. T.-Y. Li, Numerical solution of polynomial systems by homotopy continuation methods. *Handbook of Numerical Analysis*, vol. 11 (North-Holland, 2003), pp. 209–304
5. E.L. Allgower, K. Georg, *Introduction to Numerical Continuation Methods* (Society for Industrial and Applied Mathematics (SIAM), 2003)
6. A.J. Sommese, C.W. Wampler, *The Numerical Solution of Systems of Polynomials Arising in Engineering and Science* (World Scientific Publishing, 2005)
7. E.C. Sherbrooke, N.M. Patrikalakis, Computation of the solutions of nonlinear polynomial systems. Comput. Aided Geom. Des. **10**(5), 379–405 (1993)
8. E. Hansen, G.W. Walster, *Global Optimization Using Interval Analysis: Revised and Expanded*, vol. 264 of Pure and Applied Mathematics, 2nd edn. (Marcel Dekker, Inc., 2003)
9. J.M. Porta, L. Ros, F. Thomas, A linear relaxation technique for the position analysis of multiloop linkages. IEEE Trans. Robot. **25**(2), 225–239 (2009)
10. C.W. Wampler, A.J. Sommese, *21st Century Kinematics*, ch. Applying Numerical Algebraic Geometry to Kinematics (Springer, 2013), pp. 125–159
11. M.E. Henderson, *Numerical Continuation Methods for Dynamical Systems: Path Following and Boundary Value Problems*, ch. Higher-Dimensional Continuation (Springer, 2007), pp. 77–115
12. J.G. De Jalón, E. Bayo, *Kinematic and Dynamic Simulation of Multibody Systems* (Springer, 1993)
13. J.M. Porta, L. Ros, F. Thomas, F. Corcho, J. Cantó, J.J. Pérez, Complete maps of molecular-loop conformational spaces. J. Comput. Chem. **28**(13), 2170–2189 (2007)
14. O. Bohigas, M. Manubens, L. Ros, Singularities of non-redundant manipulators: a short account and a method for their computation in the planar case. Mech. Mach. Theory **68**, 1–17 (2013)
15. D. Zlatanov, *Generalized Singularity Analysis of Mechanisms*. PhD thesis, University of Toronto, 1998
16. J.K. Davidson, K.H. Hunt, *Robots and Screw Theory: Applications of Kinematics and Statics to Robotics* (Oxford University Press, 2004)

17. J. Duffy, *Statics and Kinematics with Applications to Robotics* (Cambridge University Press, 1996)
18. J.M. Porta, L. Ros, T. Creemers, F. Thomas, Box approximations of planar linkage configuration spaces. ASME J. Mech. Des. **129**(4), 397–405 (2007)
19. O. Bohigas, D. Zlatanov, L. Ros, M. Manubens, J.M. Porta, A general method for the numerical computation of manipulator singularity sets. IEEE Trans. Robot. **30**(2), 340–351 (2014)
20. N.K. Karmarkar, A new polynomial-time algorithm for linear programming. Combinatorica **4**(4), 373–395 (1984)
21. E.J. Haug, C.-M. Luh, F.A. Adkins, J.-Y. Wang, Numerical algorithms for mapping boundaries of manipulator workspaces. ASME J. Mech. Des. **118**(2), 228–234 (1996)
22. I.A. Bonev, *Geometric Analysis of Parallel Mechanisms*. PhD thesis, Faculté des Sciences et de Génie, Université de Laval, 2002
23. H. Li, C.M. Gosselin, M.J. Richard, B.M. St-Onge, Analytic form of the six-dimensional singularity locus of the general Gough-Stewart platform. ASME J. Mech. Des. **128**(1), 279–287 (2006)
24. I.A. Bonev, C.M. Gosselin, Analytical determination of the workspace of symmetrical spherical parallel mechanisms. IEEE Trans. Robot. **22**(5), 1011–1017 (2006)
25. M. Zein, P. Wenger, D. Chablat, Singular curves in the joint space and cusp points of 3-R$\underline{\text{P}}$R parallel manipulators. Robotica **25**(6), 717–724 (2007)
26. E. Macho, O. Altuzarra, C. Pinto, A. Hernandez, *Transitions Between Multiple Solutions of the Direct Kinematic Problem*, eds. by J. Lenarcic, P. Wenger. Advances in Robot Kinematics: Analysis and Design (Springer, 2008), pp. 301–310
27. P. Wenger, D. Chablat, Kinematic analysis of a class of analytic planar 3-R$\underline{\text{P}}$R parallel manipulators, in *Proceedings of the 5th International Workshop on Computational Kinematics (Duisburg, Germany)* (2009), pp. 43–50
28. The CUIK Project Home Page, http://www.iri.upc.edu/cuik. Accessed 16 Jun 2016
29. Leibniz Universität Hannover, http://www.imes.uni-hannover.de/. Accessed 16 Jun 2016
30. J. Sefrioui, C.M. Gosselin, On the quadratic nature of the singularity curves of planar three-degree-of-freedom parallel manipulators. Mech. Mach. Theory **30**(4), 533–551 (1995)
31. J.-P. Merlet, C.M. Gosselin, N. Mouly, Workspaces of planar parallel manipulators. Mech. Mach. Theory **33**(1–2), 7–20 (1998)
32. I.A. Bonev, C.M. Gosselin, Singularity loci of planar parallel manipulators with revolute joints, in *Proceedings of the 2nd Workshop on Computational Kinematics (Seoul, South Korea)* (2001), pp. 291–299
33. Companion web page of this book, http://www.iri.upc.edu/srm. Accessed 16 Jun 2016
34. B.M. St-Onge, C.M. Gosselin, Singularity analysis and representation of the general Gough-Stewart platform. Int. J. Robot. Res. **19**(3), 271–288 (2000)
35. J.-P. Merlet, *Parallel Robots* (Springer, 2006)
36. I.A. Bonev, The true origins of parallel robots (2003), http://www.parallemic.org. Accessed 26 Dec 2015
37. L.-W. Tsai, *Robot Analysis: The Mechanics of Serial and Parallel Manipulators* (Wiley-Interscience, 1999)
38. K.J. Waldron, K.H. Hunt, Series-parallel dualities in actively coordinated mechanisms. Int. J. Robot. Res. **10**(5), 473–480 (1991)
39. I.A. Bonev, D. Zlatanov, C.M. Gosselin, Advantages of the modified Euler angles in the design and control of PKMs, in *Proceedings of the 3rd Chemnitz Parallel Kinematics Seminar/2002 Parallel Kinematic Machines International Conference (Chemnitz, Germany)* (2002), pp. 171–188

Chapter 4
Workspace Determination

This chapter explains how the developments in Chaps. 2 and 3 can be extended to obtain the various workspaces of a mechanism. For a given workspace, we explain how its boundary can be found by computing a set of generalised singularities, and how the points of such set can be classified into traversable or barrier singularities. A detailed map of the workspace is obtained as a result, in which the interior and exterior regions, together with the motion impediments that separate them, become clearly identified.

The method is again general. In fact, it is equally applicable to both redundant and nonredundant mechanisms because the shape of the workspace depends only on the underlying mechanism and not on the specific actuation pattern adopted. After reviewing the state of the art on the topic (Sect. 4.1), we describe the notion of workspace mathematically, and then derive the conditions allowing the detection of the boundary points, and the traversable and barrier singularities (Sect. 4.2). We then explain the main limitations of a continuation approach of a similar generality (Sect. 4.3), and the alternative method we propose to overcome them (Sect. 4.4). We finally apply the method to several mechanisms that would be difficult to analyse using the continuation approach (Sect. 4.5).

4.1 The Need of a General Method

Since the early studies on the problem [1–5], efficient workspace determination methods have been given, but most of them are tailored to a particular robot architecture or narrow class of architectures. An important group of methods adopts a constructive geometric approach. Representative of them is [6], which computes the constant-orientation workspace of a spatial parallel mechanism; [7], which extends the approach to deal with other physical constraints; or [8], which provides methods for various planar parallel platforms. Other significant approaches include interval analysis or discretisation techniques for Gough-Stewart platforms [9, 10], optimisation-based algorithms for fully serial or parallel robots

O. Bohigas et al., *Singularities of Robot Mechanisms*,
Mechanisms and Machine Science 41, DOI 10.1007/978-3-319-32922-2_4

[11], analytic methods for symmetrical spherical mechanisms [12], and analytic, topologic, or algebraic-geometric procedures for serial mechanisms [13–17].

The literature about the topic is extensiven (see [11, 18–20] for elaborate surveys), but most previous methods are difficult to apply to mechanisms outside the class they consider because they exploit specificities of the class in some way or another, like the possibility to parameterise the end-effector pose [13, 21], simplifications introduced by dimension symmetries [12], or the feasibility of an algebraic or geometric treatment [6–8, 13–17]. Even discretisation methods, which resort to configuration sampling only [5, 10, 22], rely on the assumption that either the forward or the inverse kinematic problems have a simple solution, which needs not be the case in general. Section 4.5.4 shows an illustrative example on this regard.

While *ad-hoc* methods are desirable because they tend to yield faster or simpler algorithms, methods for general architectures are required too, to be able to analyse robots for which no specific solution exists. Up to our knowledge, the only approach that exhibits a degree of generality close to that of our method is the one due to Haug and collaborators (see [23] and references therein). In their work, Haug et al. apply the idea that the workspace boundaries can be extracted from a set of generalised singularities, and focus their effort on tracing such singularities numerically using a continuation method. The procedure is elegant and it can be applied to arbitrary lower-pair mechanisms, but important weaknesses were identified in [24], including (1) the need to manually guide the method with a priori knowledge of the workspace shape, (2) the fact that only cross-section curves of the boundary can be obtained in higher-dimensional cases, or (3) the possibility to miss some boundary segments when voids are present inside the workspace. Unfortunately, the performance of the procedure can further degrade, since one can encounter workspaces with several connected components, hidden barriers, or degenerate barriers where the method will produce incomplete or misleading maps of the workspace (Sect. 4.3).

4.2 The Workspace and Its Boundaries

The motion capabilities of a mechanism are typically assessed by analysing the set of feasible poses of the end-effector, which is usually known as the *workspace* or the *accessible output set* of the mechanism. However, the notion of workspace is more general. The regions swept by other configuration coordinates could equally be useful to the robot designer, including those of the input coordinates, which identify the motion ranges of the actuators, or those of a point on any link, which could reveal possible collisions with the environment. In this chapter, thus, we shall be interested in computing the set of attainable values of a generic tuple $\mathbf{u}$ of n_u coordinates, as the mechanism runs through all possible configurations. Note that such a tuple will only coincide with the homonym tuple of Chap. 2 when computing the workspace of the output coordinates. Moreover, since joint mechanical limits can greatly modify the workspace shape, we shall include them in the formulation of Eq. (2.1), as explained in Sect. 4.4.1. Therefore, the C-space of the mechanism will only contain points $\mathbf{q}$ fulfilling such limits.

Depending on the choice, the $\mathbf{u}$ coordinates may be explicit in $\mathbf{q}$ or not. However, we shall show that it is always possible to derive a system of the form of Eq. (2.1) containing the $\mathbf{u}$ variables of interest, so that $\mathbf{q}$ will be assumed to include such variables hereafter. By adopting the partition $\mathbf{q} = (\mathbf{z}, \mathbf{u})$, where $\mathbf{z}$ accumulates all coordinates in $\mathbf{q}$ except those in $\mathbf{u}$, Eq. (2.1) can then be written as

$$\mathbf{\Phi}(\mathbf{z}, \mathbf{u}) = \mathbf{0}, \tag{4.1}$$

and the workspace of the system relative to the $\mathbf{u}$ coordinates can be defined as the set $\mathcal{A}$ of points $\mathbf{u} \in \mathcal{U}$ that satisfy Eq. (4.1) for some value of $\mathbf{z}$, where $\mathcal{U}$ is here the total n_u-dimensional space where the $\mathbf{u}$ tuple takes values. We shall also assume that n_u is less than or equal to the dimension n of the C-space, and that $\mathcal{A}$ is a subset of $\mathcal{U}$ of dimension n_u, which is the common situation in practice. In particular, if n_z is the size of $\mathbf{z}$, and n_e is the number of equations in Eq. (2.1), $n_u \leq n$ implies that $n_z \geq n_e$, so that the system in Eq. (4.1) is in general not overconstrained for a fixed value of $\mathbf{u}$.

Although a direct computation of $\mathcal{A}$ can always be attempted (e.g., by solving Eq. (4.1) with the branch-and-prune methods in Chap. 3), we shall study the boundary of $\mathcal{A}$ instead, which will result in richer information on the motion impediments to be found in $\mathcal{U}$ by the mechanism. A point $\mathbf{u}$ lies in the boundary of $\mathcal{A}$ if every open set of $\mathcal{U}$ containing $\mathbf{u}$ intersects the interior and the exterior of $\mathcal{A}$. The boundary of $\mathcal{A}$ will be denoted by $\partial\mathcal{A}$.

4.2.1 Jacobian Rank Deficiency Conditions

To derive the conditions allowing the identification of $\partial\mathcal{A}$, observe first that the workspace $\mathcal{A}$ is exactly the image of $\mathcal{C}$ through $\pi_{\mathbf{u}}$, the map projecting $\mathbf{q}$ to the $\mathbf{u}$ coordinates:

$$\pi_{\mathbf{u}}(\mathbf{z}, \mathbf{u}) = \mathbf{u}.$$

It is not difficult to see, moreover, that the Jacobian matrix

$$\mathbf{\Phi_z}(\mathbf{q}) = \begin{bmatrix} \frac{\partial \Phi_1}{\partial z_1} & \cdots & \frac{\partial \Phi_1}{\partial z_{n_z}} \\ \vdots & & \vdots \\ \frac{\partial \Phi_{n_e}}{\partial z_1} & \cdots & \frac{\partial \Phi_{n_e}}{\partial z_{n_z}} \end{bmatrix}$$

must be rank deficient at the points $\mathbf{q} \in \mathcal{C}$ projecting to some $\mathbf{u} \in \partial\mathcal{A}$. Certainly, note that if $\mathbf{\Phi_z}$ is full rank at a point $\mathbf{q} = (\mathbf{z}, \mathbf{u}) \in \mathcal{C}$, then there exists a nonvanishing

$n_e \times n_e$ minor of $\mathbf{\Phi_z}$, say relative to some variables $\mathbf{z}'$, and by the implicit function theorem it is possible to find a function $\mathbf{z}' = \mathbf{F}(\mathbf{u}')$ relating $\mathbf{z}'$ with the remaining $\mathbf{u}'$ variables, satisfying

$$\mathbf{\Phi}(\mathbf{F}(\mathbf{u}'), \mathbf{u}') = \mathbf{0}.$$

Thus, the $\mathbf{u}'$ variables, which include the $\mathbf{u}$ ones, can be used as a local parameterisation of $\mathcal{C}$ around $(\mathbf{z}', \mathbf{u}')$. This implies that arbitrary values in a neighbourhood of a point $\mathbf{u} \in \mathcal{U}$ have a corresponding $\mathbf{z}$ satisfying $\mathbf{\Phi}(\mathbf{z}, \mathbf{u}) = \mathbf{0}$, so that $\mathbf{u}$ must lie in the interior of $\mathcal{A}$. It is clear, then, that $\mathbf{\Phi_z}$ should be rank deficient for $\mathbf{u}$ to belong to $\partial\mathcal{A}$.

Geometrically, the points $\mathbf{q} \in \mathcal{C}$ where $\mathbf{\Phi_z}$ is rank deficient are the critical points of the projection of $\mathcal{C}$ to $\mathcal{U}$, i.e. those in which the projection of the tangent space of $\mathcal{C}$ does not span the tangent space of $\mathcal{U}$ at $\mathbf{u} = \pi_\mathbf{u}(\mathbf{q})$, so that the mechanism, as a result, loses instantaneous mobility (i.e., dexterity) relative to the $\mathbf{u}$ variables. Since this result is quite similar to the one obtained for the forward and inverse singularities in Sect. 2.2.3, the set $\mathcal{W}$ of all critical points of the projection of $\mathcal{C}$ to $\mathcal{U}$ will be called the *singularity set* in this chapter. When $\mathbf{u}$ coincides with the input or output coordinates, the set $\mathcal{W}$ will include the forward or inverse singularities, respectively, but also other points corresponding to the loss of mobility introduced by mechanical limits at the joints (Sect. 4.4.1).

In sum, we can gain a preliminary idea of how the workspace boundary would look like by computing $\mathcal{W}$ and projecting it down to the $\mathbf{u}$ variables, obtaining $\pi_\mathbf{u}(\mathcal{W})$ as a result. Figure 4.1a illustrates the process when $\mathcal{Q}$ is $\mathbb{R}^3$, $\mathcal{C}$ is the sphere

$$x^2 + y^2 + z^2 = 1,$$

$\mathcal{U} = \mathbb{R}^2$, and $\pi_\mathbf{u}$ is the projection map $f(x, y, z) = (x, y)$. The workspace relative to the (x, y) coordinates is the projection of the sphere to the (x, y) plane, and the

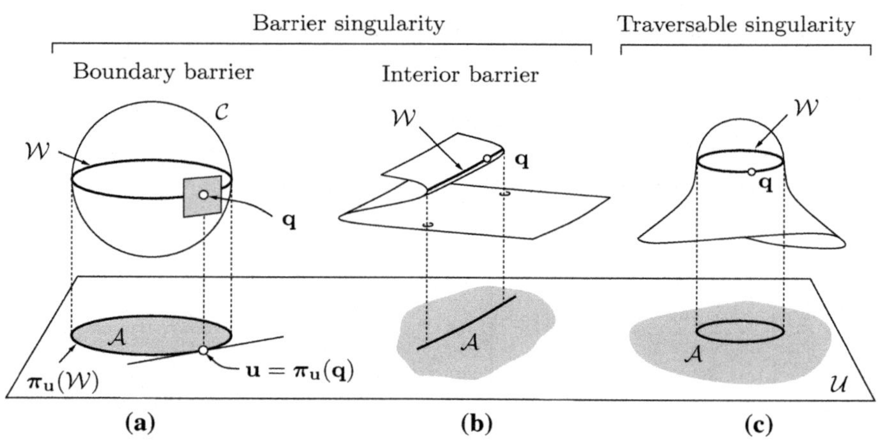

Fig. 4.1 Boundary barriers, interior barriers, and traversable singularities

boundaries of such projection necessarily correspond to points on the sphere where the tangent plane projects to a line of $\mathbb{R}^2$.

Note from the examples in Figs. 4.1b and c, however, that the rank deficiency of $\mathbf{\Phi_z}$ is a necessary, but not a sufficient condition for $\pi_{\mathbf{u}}(\mathbf{q})$ to lie in $\partial\mathcal{A}$, as there can be critical points projecting to the interior of $\mathcal{A}$ too. We next provide a way to classify the points of $\mathcal{W}$ according to whether they actually correspond to motion impediments in the workspace.

4.2.2 Barrier Analysis

Figure 4.1 illustrates that points $\mathbf{q}$ where $\mathbf{\Phi_z}$ is rank deficient can be classified into two broad categories. They can be *traversable* or *barrier* singularities depending on whether there exists a trajectory in $\mathcal{C}$ through $\mathbf{q}$ whose projection on $\mathcal{U}$ traverses $\pi_{\mathbf{u}}(\mathcal{W})$ for each neighbourhood of $\mathbf{q}$ in $\mathcal{C}$ [19, 25]. Points corresponding to barrier singularities can in turn be classified into *boundary* or *interior* barriers according to whether they occur over $\partial\mathcal{A}$, or over the interior of $\mathcal{A}$, respectively. An example of each one of these singularities is depicted in Fig. 4.2 for a planar 3R robot arm.

As noted in [23–25], a workspace determination method should ideally detect *all* barriers in the workspace, either interior or on the boundary, because such barriers constitute true motion impediments with respect to the $\mathbf{u}$ variables. A criterion to

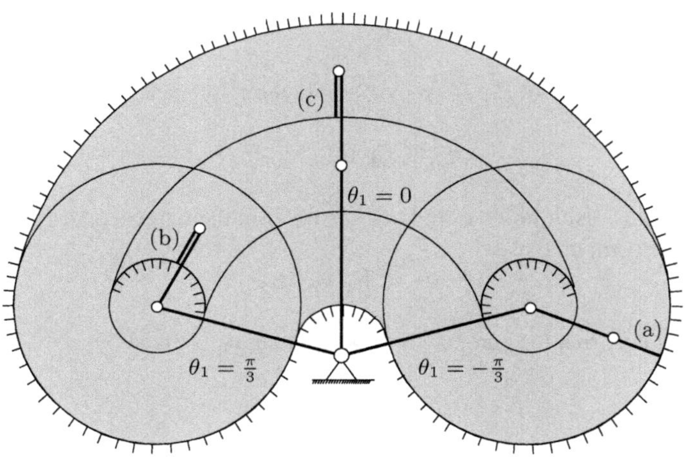

Fig. 4.2 Workspace of a planar 3R arm mechanism relative to the (x, y) coordinates of the tip point of the last link, assuming that the angle θ_1 of the first revolute joint is restricted to the $[-\pi/3, \pi/3]$ range. Points corresponding to singularities are indicated in *solid lines*, and those relative to boundary and interior barriers are indicated with normal vectors in the forbidden side. Configurations a, b, and c correspond to a boundary barrier, an interior barrier, and a traversable singularity, respectively

determine whether a point $\mathbf{q}_0 = (\mathbf{z}_0, \mathbf{u}_0) \in \mathcal{W}$ corresponds to a barrier or a traversable singularity was given in [23]. We next recall its main points.

Suppose that $\mathbf{q} = \mathbf{q}(\mathbf{v})$ is a smooth parameterisation of $\mathcal{C}$ in a neighbourhood of $\mathbf{q}_0 \in \mathcal{W}$, where $\boldsymbol{\Phi}_{\mathbf{q}}(\mathbf{q}_0)$ is full rank, $\mathbf{v}$ is a tuple of n parameters, and $\mathbf{q}_0 = \mathbf{q}(\mathbf{v}_0)$ for some $\mathbf{v}_0 \in \mathbb{R}^n$. The implicit function theorem guarantees that such a parameterisation exists because $\boldsymbol{\Phi}_{\mathbf{q}}(\mathbf{q}_0)$ is full rank [26]. Let also $\mathbf{n}_0$ be a normal vector to $\boldsymbol{\pi}_{\mathbf{u}}(\mathcal{W})$ at $\mathbf{u}_0$. We can determine whether $\mathbf{q}_0$ corresponds to a barrier singularity by studying the sign of the function

$$\chi(\mathbf{v}) = \mathbf{n}_0^{\mathsf{T}}(\mathbf{u}(\mathbf{v}) - \mathbf{u}_0) \tag{4.2}$$

for all local trajectories $\mathbf{v} = \mathbf{v}(t)$ crossing $\mathbf{v}_0$ for some $t = t_0$, and whose corresponding path $\mathbf{u} = \mathbf{u}(t)$ is orthogonal to $\boldsymbol{\pi}_{\mathbf{u}}(\mathcal{W})$ at $\mathbf{u}_0$.

To perform such study, we first compute $\mathbf{n}_0$ as follows. Assume that

$$\mathbf{q}(t) = (\mathbf{z}(t), \mathbf{u}(t)) \tag{4.3}$$

is an arbitrary trajectory in $\mathcal{C}$ parameterised in t, traversing $\mathbf{q}_0 = (\mathbf{z}_0, \mathbf{u}_0)$ for $t = t_0$. By substituting Eq. (4.3) into Eq. (4.1) and evaluating the time derivatives of this equation at $\mathbf{q} = \mathbf{q}_0$, we get

$$\boldsymbol{\Phi}_{\mathbf{u}}(\mathbf{q}_0)\dot{\mathbf{u}}(t_0) + \boldsymbol{\Phi}_{\mathbf{z}}(\mathbf{q}_0)\dot{\mathbf{z}}(t_0) = \mathbf{0}. \tag{4.4}$$

Since $\mathbf{q}_0 \in \mathcal{W}$, $\boldsymbol{\Phi}_{\mathbf{z}}(\mathbf{q}_0)$ must be rank deficient, and thus there must exist a nonnull vector $\boldsymbol{\xi}_0$ such that

$$\boldsymbol{\Phi}_{\mathbf{z}}(\mathbf{q}_0)^{\mathsf{T}}\boldsymbol{\xi}_0 = \mathbf{0}. \tag{4.5}$$

Multiplying both sides of Eq. (4.4) by $\boldsymbol{\xi}_0^{\mathsf{T}}$ and taking into account Eq. (4.5), we get

$$\boldsymbol{\xi}_0^{\mathsf{T}}\boldsymbol{\Phi}_{\mathbf{u}}(\mathbf{q}_0)\dot{\mathbf{u}}(t_0) = 0, \tag{4.6}$$

and because this result holds for all vectors $\dot{\mathbf{u}}(t_0)$ in the tangent space of $\boldsymbol{\pi}_{\mathbf{u}}(\mathcal{W})$ at $\mathbf{u}_0$, $\mathbf{n}_0$ can be computed as

$$\mathbf{n}_0 = \boldsymbol{\Phi}_{\mathbf{u}}(\mathbf{q}_0)^{\mathsf{T}}\boldsymbol{\xi}_0. \tag{4.7}$$

Incidentally, note that the component of $\dot{\mathbf{u}}(t)$ along $\mathbf{n}_0$ is given by

$$\dot{\mathbf{u}}_{\mathbf{n}_0} = \mathbf{n}_0^{\mathsf{T}}\dot{\mathbf{u}}(t_0), \tag{4.8}$$

but using Eqs. (4.6) and (4.7) we realise that

$$\dot{\mathbf{u}}_{\mathbf{n}_0} = \boldsymbol{\xi}_0^{\mathsf{T}}\boldsymbol{\Phi}_{\mathbf{u}}(\mathbf{q}_0)\dot{\mathbf{u}}(t_0) = 0 \tag{4.9}$$

independently of the chosen trajectory $\mathbf{q}(t)$ in $\mathcal{C}$ through $\mathbf{q}_0$. Thus, $\dot{\mathbf{u}}(t_0)$ must necessarily lie in the tangent space of $\boldsymbol{\pi}_{\mathbf{u}}(\mathcal{W})$ at $\mathbf{u}_0$, which shows that there is a loss of mobility/dexterity relative to the $\mathbf{u}$ variables at all points of $\mathcal{W}$.

The sign of $\chi(\mathbf{v})$ in Eq. (4.2) can now be studied by resorting to the second-order Taylor expansion of $\chi(\mathbf{v})$ around $\mathbf{v}_0$

$$\chi(\mathbf{v}) \simeq \chi(\mathbf{v}_0) + \delta\mathbf{v}^{\mathsf{T}}\chi_{\mathbf{v}}(\mathbf{v}_0) + \frac{1}{2}\delta\mathbf{v}^{\mathsf{T}}\chi_{\mathbf{vv}}(\mathbf{v}_0)\delta\mathbf{v},$$

where $\chi_{\mathbf{v}}$ and $\chi_{\mathbf{vv}}$ are the gradient and Hessian of $\chi(\mathbf{v})$, and $\delta\mathbf{v} = (\mathbf{v} - \mathbf{v}_0)$ is a small displacement whose corresponding $\delta\mathbf{u} = (\mathbf{u} - \mathbf{u}_0)$ is orthogonal to $\boldsymbol{\pi}_{\mathbf{u}}(\mathcal{W})$. Note that the first term of the previous expansion vanishes because

$$\chi(\mathbf{v}_0) = \mathbf{n}_0^{\mathsf{T}}(\mathbf{u}_0 - \mathbf{u}_0) = 0.$$

Moreover, the time derivative of Eq. (4.2) for $\mathbf{v} = \mathbf{v}(t)$ is

$$\dot{\chi}(t) = \mathbf{n}_0^{\mathsf{T}}\dot{\mathbf{u}}(t),$$

which, for $t = t_0$ is null according to Eqs. (4.8) and (4.9). Since for $t = t_0$ we have $\dot{\chi} = \chi_{\mathbf{v}}\dot{\mathbf{v}} = 0$ for all $\dot{\mathbf{v}}$, we conclude that $\chi_{\mathbf{v}}(\mathbf{v}_0) = 0$ as well, meaning that the second term of the Taylor expansion also vanishes.

In conclusion, the sign of $\chi(\mathbf{v})$ is mostly determined by the definiteness properties of the quadratic form $\delta\mathbf{v}^{\mathsf{T}}\chi_{\mathbf{vv}}(\mathbf{v}_0)\delta\mathbf{v}$. If this form is positive- or negative-definite, then all trajectories orthogonal to $\boldsymbol{\pi}_{\mathbf{u}}(\mathcal{W})$ lie in one side of $\boldsymbol{\pi}_{\mathbf{u}}(\mathcal{W})$ and $\mathbf{q}_0$ is a barrier singularity. If this form is indefinite, there are trajectories in $\mathcal{A}$ that cross $\boldsymbol{\pi}_{\mathbf{u}}(\mathcal{W})$ and $\mathbf{q}_0$ is a traversable singularity. Lastly, if this form is semi-definite, we cannot deduce the singularity type unless we examine higher-order terms of the Taylor expansion. However, the latter case only occurs in zero-measure subsets of $\mathcal{W}$ in general.

The definiteness test just outlined can be implemented by checking the eigenvalues of $\chi_{\mathbf{vv}}(\mathbf{v}_0)$, as explained in [23].

4.3 Issues of Continuation Methods

In order to see the advantages of our approach in comparison to the continuation method in [23], this method is next reviewed briefly, identifying a number of cases where it fails to produce complete maps of the workspace.

We note first that the method in [23] relies on one-dimensional path-tracking procedures, and hence it can only trace $\partial\mathcal{A}$ explicitly on one-dimensional boundaries, i.e. when $n_u = 2$. Figure 4.3a explains the method in a simple setting in which $\mathcal{Q} = \mathbb{R}^3$, $\mathcal{U} = \mathbb{R}^2$, and $\mathcal{C}$ contains two connected components, $\mathcal{C}_1$ and $\mathcal{C}_2$, which project to a workspace $\mathcal{A}$ with two regions.

The method begins with a manually-provided configuration of the mechanism, $\mathbf{q}_i = (\mathbf{z}_i, \mathbf{u}_i)$, with $\mathbf{u}_i \in \mathcal{A}$. Then it shoots a ray through $\mathbf{u}_i$ in $\mathcal{U}$ along an arbitrary direction, and traces this ray until a point of a barrier singularity is found. The movement along this ray is achieved by continuation of the corresponding path in

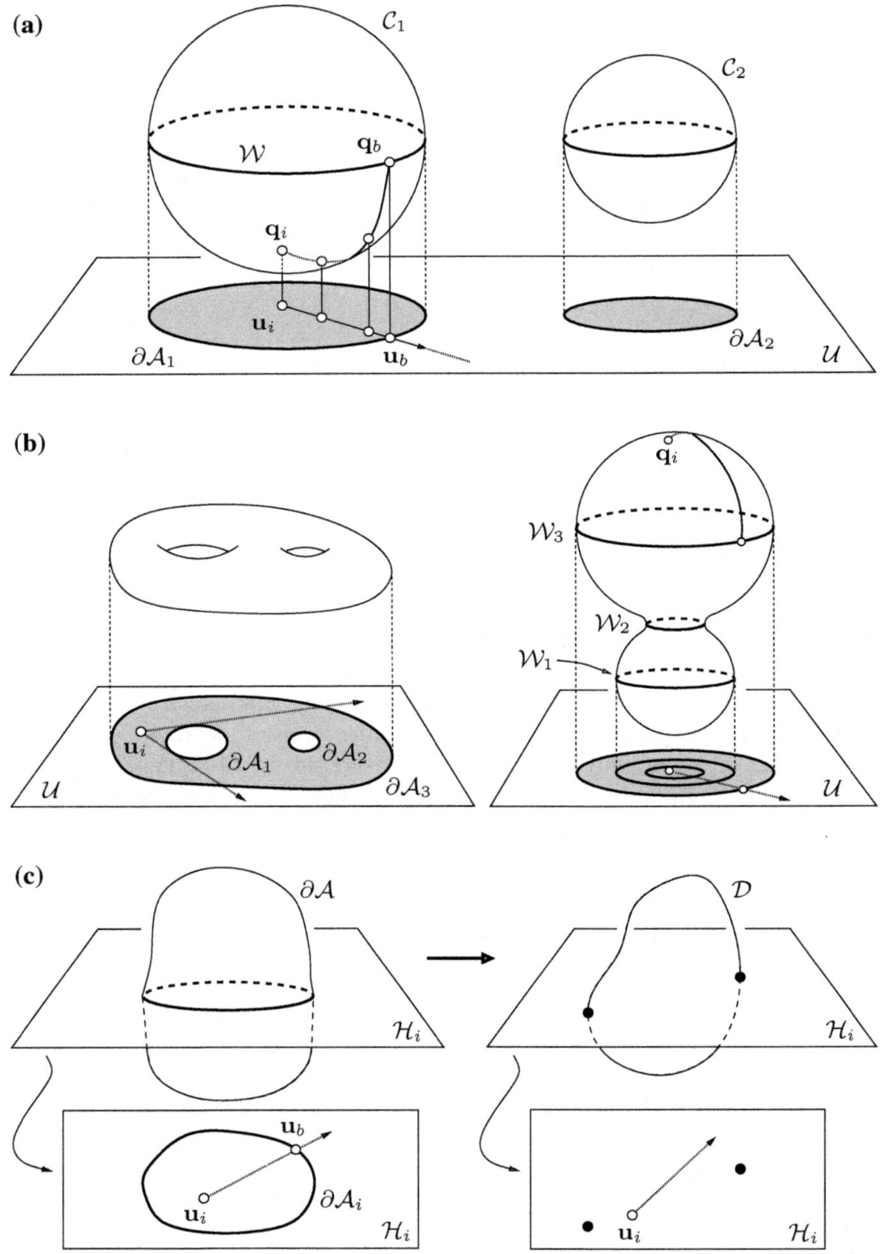

Fig. 4.3 Performance of the continuation method in multicomponent workspaces (**a**), hidden barriers (**b**), and degenerate barriers (**c**)

$\mathcal{C}$, i.e. by iteratively solving $\mathbf{\Phi}(\mathbf{z}, \mathbf{u}) = \mathbf{0}$ using a Newton-Raphson method for $\mathbf{u}$ fixed to discrete values along the ray. This process is repeated until a value of $\mathbf{u}$ outside $\mathcal{A}$ is found, which is detected because the Newton-Raphson method fails to converge. A dichotomy-search process is then performed locally to find a point $\mathbf{q}_b \in \mathcal{C}$ lying in $\mathcal{W}$. A second continuation process is then launched from $\mathbf{q}_b$ to find the connected component of $\mathcal{W}$ that is reachable from such point, by solving a system of equations that expresses the rank deficiency of $\mathbf{\Phi}_{\mathbf{z}}$. Once $\mathcal{W}$ has been traced, the points of $\pi_{\mathbf{u}}(\mathcal{W})$ are computed by projection, and those corresponding to barrier or traversable singularities are finally detected by means of the method outlined in the previous section.

Because path-tracking methods based on continuation are fast and robust to bifurcations [27], this approach will rapidly determine $\mathcal{W}$ in favorable cases. However, one encounters the following scenarios in which the method will fail to identify $\partial\mathcal{A}$ completely.

Multicomponent workspaces

Difficulties arise, for example, when computing $\partial\mathcal{A}$ in workspaces with several connected components, like the one in Fig. 4.3a. Independently of the chosen direction for the ray, note that the previous process will certainly hit $\partial\mathcal{A}_1$, but not $\partial\mathcal{A}_2$, because the tracking of the ray cannot continue beyond $\partial\mathcal{A}_1$ using continuation. To converge to all boundary curves, the previous strategy should at least be fed with one point of each connected component of $\mathcal{C}$, but no satisfactory method has been given yet to compute such points in general, to the best of our knowledge.

Hidden barriers

The continuation method may seem to be able to compute, at least, all barriers of the workspace component to which $\mathbf{q}_i$ belongs, but this is not always the case. Note that $\mathcal{W}$ may itself have several connected components, and some of such components could be missed depending on the position of $\mathbf{u}_i$, even if rays along all possible directions were shot. In Fig. 4.3b, left, for example, the continuation method may be able to find $\partial\mathcal{A}_1$ and $\partial\mathcal{A}_3$ from $\mathbf{u}_i$, but not $\partial\mathcal{A}_2$, because $\partial\mathcal{A}_2$ is hidden behind $\partial\mathcal{A}_1$. The problem also arises in workspaces with interior barriers, as seen in Fig. 4.3b, right. Starting the continuation from $\mathbf{q}_i$ allows us to hit the boundary barrier corresponding to $\mathcal{W}_3$, but not the interior barriers corresponding to $\mathcal{W}_1$ and $\mathcal{W}_2$, thus ignoring true obstacles lying inside the workspace.

Degenerate barriers

When $\partial\mathcal{A}$ has dimension larger than one, the method in [23] slices $\partial\mathcal{A}$ through hyperplanes $\mathcal{H}_i$, so as to obtain one-dimensional curves $\partial\mathcal{A}_i$ traceable by one-dimensional continuation (Fig. 4.3c, left). In mechanisms of a special geometry, however, portions of $\partial\mathcal{A}$ can degenerate into lower-dimensional barriers $\mathcal{D}$, making the slices $\mathcal{H}_i$ contain isolated points only (Fig. 4.3c, right). The method will clearly miss the lower-dimensional barriers in such a situation, because the ray shooting technique will fail to meet them with probability one, independently of the location of $\mathbf{u}_i$.

As we shall see in Sect. 4.5, examples of multicomponent workspaces, hidden barriers, and degenerate barriers occur easily in real mechanisms, which calls for the development of alternative methods that are robust to such situations. We next propose one based on the results of Sect. 4.2.

4.4 Exploiting the Branch-and-Prune Machinery

The method we propose exploits the same numerical approach of Chap. 3. It consists in modelling the joint limits as equality constraints (Sect. 4.4.1), then deriving a system of quadratic equations characterising the set $\mathcal{W}$ under consideration (Sect. 4.4.2), and finally computing a box approximation of $\mathcal{W}$ of sufficient resolution in order to classify the points of such set into boundary barriers, interior barriers, or traversable singularities (Sect. 4.4.3).

4.4.1 Joint Limit Constraints

Although mechanical limits at the joints are usually described by means of inequalities, we shall take them into account using equality constraints, which allows us to directly apply the Jacobian rank defficiency conditions of Sect. 4.2.1 to obtain the boundary of the workspace. Two types of limits need to be treated: those referring to angular constraints and those referring to distance constraints. These are needed to model revolute and prismatic joints, respectively, but combinations of both limits may be encountered in other joint types.

Following the formulation of Sect. 3.2, let L_j and L_k be two links connected through a revolute joint J_i. The relative angle ϕ_i between L_j and L_k is defined as the angle between two unit vectors $\mathbf{k}_i$ and $\mathbf{n}_i$ orthogonal to the axis of J_i, moving rigidly with L_j and L_k respectively. Then, suppose that we want to limit ϕ_i to lie in the interval

$$\phi_i \in [-\phi_i^l, \phi_i^l].$$

Instead of introducing ϕ_i in the formulation, which would be impossible using quadratic equations, we shall constrain the limits of ϕ_i indirectly by setting $\cos\phi_i \geq \cos\phi_i^l$. Note for this matter that if $c_i = \cos(\phi_i)$ then c_i is related to $\mathbf{k}_i$ and $\mathbf{n}_i$ through

$$c_i = \mathbf{k}_i^\mathsf{T}\mathbf{n}_i. \tag{4.10}$$

Moreover, the fact that $\mathbf{k}_i$ and $\mathbf{n}_i$ are fixed in $\mathcal{F}_j$ and $\mathcal{F}_k$ implies that

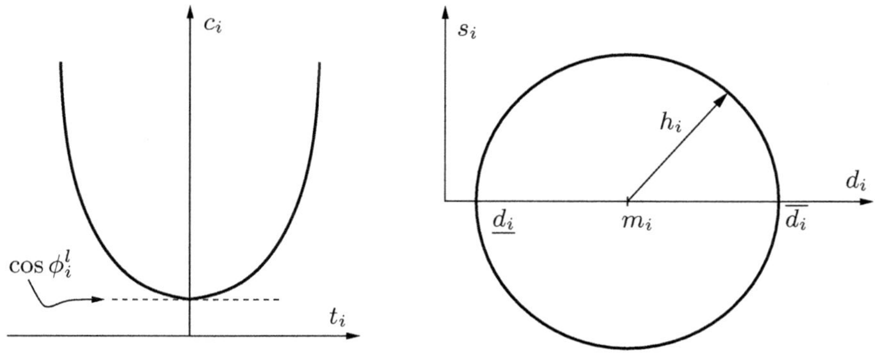

Fig. 4.4 Auxiliary constraints to enforce joint angle and distance limits

$$\mathbf{k}_i = \mathbf{R}_j\,\mathbf{k}_i^{\mathcal{F}_j}, \tag{4.11}$$

$$\mathbf{n}_i = \mathbf{R}_k\,\mathbf{n}_i^{\mathcal{F}_k}. \tag{4.12}$$

Thus, to limit ϕ_i we can simply add Eqs. (4.10)–(4.12) to the system to be solved, together with

$$c_i = t_i^2 + \cos\phi_i^l, \tag{4.13}$$

where t_i is a new auxiliary variable that can take any value (Fig. 4.4, left). Observe that the constraint $\phi_i \in [-\phi_i^l, \phi_i^l]$ is satisfied if, and only if, Eq. (4.13) is satisfied for some value of t_i.

The same equations can be used to limit the angle in universal and spherical joints, where the vectors $\mathbf{k}_i$ and $\mathbf{n}_i$ can be choosen as shown in Fig. 4.5.

If L_j and L_k are connected by means of a prismatic joint J_i, the displacement d_i of the joint is already present in the system as a variable. Suppose that, in this context, we want to constrain d_i to lie within the interval

$$d_i \in \left[\,\underline{d_i}, \overline{d_i}\,\right].$$

To do so, we can readily impose this constraint by setting

$$(d_i - m_i)^2 + s_i^2 = h_i^2, \tag{4.14}$$

where s_i is a newly-defined auxiliary variable, and

$$m_i = \frac{\overline{d_i} + \underline{d_i}}{2}, \qquad h_i = \frac{\overline{d_i} - \underline{d_i}}{2}.$$

The values m_i and h_i are called the *mid-point* and *half-range* of the interval $\left[\,\underline{d_i}, \overline{d_i}\,\right]$, and Eq. (4.14) simply constrains the pairs (d_i, s_i) to take values on a circle of radius h_i

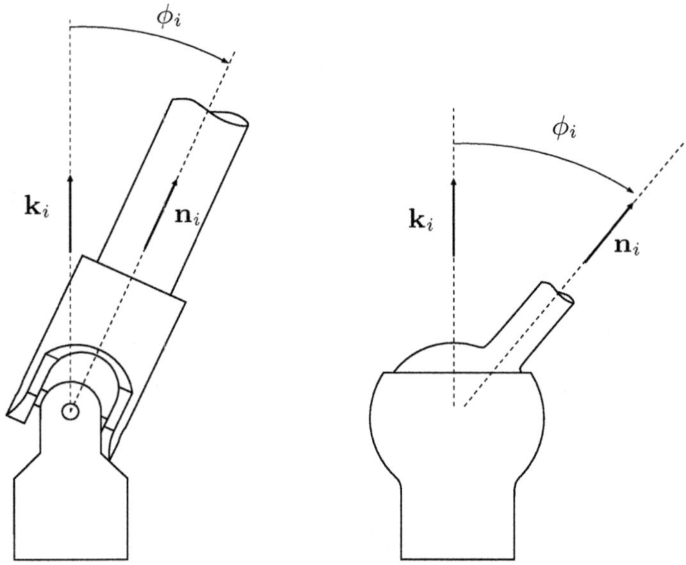

Fig. 4.5 Vectors intervening in mechanical limits of universal and spherical joints

centred at $(m_i, 0)$ in the (d_i, s_i) plane (Fig. 4.4, right). As a consequence, $d_i \in [\underline{d_i}, \overline{d_i}]$ if, and only if, Eq. (4.14) is satisfied for some s_i.

If we introduce the equations that model the joint limits into Eq. (2.1), the variables t_i and s_i become part of $\mathbf{q}$, and we obtain a new system of equations whose solution set contains the feasible configurations according to such limits.

4.4.2 Equations of the Generalised Singularity Set

To obtain a system of equations characterising $\mathcal{W}$, we shall distinguish two situations, depending on whether the $\mathbf{u}$ variables appear explicitly in $\mathbf{q}$, or are only determined implicitly by some of the variables in $\mathbf{q}$. Both situations are illustrated in Sect. 4.5 with particular examples. In all cases it will be assumed that the assembly constraints in Eq. (2.1) have been formulated following Sect. 3.2, where the pose of the links appear as $(\mathbf{r}_j, \mathbf{R}_j)$, where $\mathbf{r}_j$ and $\mathbf{R}_j$ respectively encompass the Cartesian coordinates of a point on the link, and the components of a rotation matrix giving the orientation of the link. Note that the equations added to model the joint limits are quadratic, so the whole formulation of Eq. (2.1) remains quadratic too.

Dealing with Explicit Workspace Variables

Assume initially that $\mathbf{q}$ explicitly contains $\mathbf{u}$. This occurs usually in positional workspaces, i.e. when $\mathbf{u}$ includes part or all of the variables of some $\mathbf{r}_j$. By adopting the partition $\mathbf{q} = (\mathbf{z}, \mathbf{u})$, $\mathcal{W}$ is the set of points satisfying $\boldsymbol{\Phi}(\mathbf{z}, \mathbf{u}) = \mathbf{0}$ for which $\boldsymbol{\Phi}_{\mathbf{z}}$ is rank deficient, i.e. the set of points $\mathbf{q}$ that satisfy

$$\left.\begin{array}{r}\boldsymbol{\Phi}(\mathbf{z}, \mathbf{u}) = \mathbf{0} \\ \boldsymbol{\Phi}_{\mathbf{z}}^{\mathsf{T}} \boldsymbol{\xi} = \mathbf{0} \\ \boldsymbol{\xi}^{\mathsf{T}} \boldsymbol{\xi} = 1\end{array}\right\}, \tag{4.15}$$

for some $\boldsymbol{\xi}$, where $\boldsymbol{\xi}$ is an n_e-dimensional vector of unknowns, and $\boldsymbol{\Phi}_{\mathbf{z}}$ is transposed since, if $n_u < n$, it is a rectangular matrix with more columns than rows. Clearly, the first equation in (4.15) constrains $\mathbf{q}$ to be a valid configuration respecting the joint limits, and the second and third equations impose the rank deficiency of $\boldsymbol{\Phi}_{\mathbf{z}}$. Since Eq. (2.1) is quadratic, Eq. (4.15) will be quadratic too, because all entries in $\boldsymbol{\Phi}_{\mathbf{z}}$ will be linear terms, and $\boldsymbol{\xi}^{\mathsf{T}} \boldsymbol{\xi}$ is directly a quadratic expression. Thus, a system characterising $\mathcal{W}$ adopts the form of Eq. (4.15) in this case.

Dealing with Implicit Workspace Variables

There are situations in which the chosen $\mathbf{u}$ variables do not all intervene in $\mathbf{q}$ but, instead, they can be related to a subset $\tilde{\mathbf{u}}$ of $n_{\tilde{u}}$ variables in $\mathbf{q}$, through a smooth function of the form

$$\tilde{\mathbf{u}} = \boldsymbol{\mu}(\mathbf{u}). \tag{4.16}$$

This occurs whenever $\mathbf{u}$ contains orientation angles of some link, for instance. Since the formulation of the assembly constraints represents orientations by rotation matrices (Sect. 3.2.1), orientation angles are only related implicitly to the components of $\mathbf{R}_j$ through a parameterisation of the special orthogonal group under consideration ($SO(2)$ or $SO(3)$ depending on whether the mechanism is planar or spatial). In order to transform Eq. (2.1) into the form of Eq. (4.1) it will be possible, in such situations, to consider the following partition

$$\mathbf{q} = (\tilde{\mathbf{z}}, \tilde{\mathbf{u}})$$

and subdivide Eq. (2.1) into the two subsystems

$$\left.\begin{array}{r}\boldsymbol{\Psi}(\tilde{\mathbf{z}}, \tilde{\mathbf{u}}) = \mathbf{0} \\ \boldsymbol{\eta}(\tilde{\mathbf{u}}) = \mathbf{0}\end{array}\right\}, \tag{4.17}$$

where $\boldsymbol{\eta}(\tilde{\mathbf{u}}) = \mathbf{0}$ is a subsystem of equations whose solution set can be globally parameterised by Eq. (4.16), and $\boldsymbol{\Psi}(\tilde{\mathbf{z}}, \tilde{\mathbf{u}}) = \mathbf{0}$ collects the rest of equations. Since $\tilde{\mathbf{u}} = \boldsymbol{\mu}(\mathbf{u})$ parameterises the solution set of $\boldsymbol{\eta}(\tilde{\mathbf{u}}) = \mathbf{0}$, $\boldsymbol{\eta}(\tilde{\mathbf{u}}) = \mathbf{0}$ can be replaced by $\tilde{\mathbf{u}} = \boldsymbol{\mu}(\mathbf{u})$ in Eq. (4.17), obtaining the equivalent system

$$\left.\begin{array}{r}\boldsymbol{\Psi}(\tilde{\mathbf{z}}, \tilde{\mathbf{u}}) = \mathbf{0} \\ \tilde{\mathbf{u}} = \boldsymbol{\mu}(\mathbf{u})\end{array}\right\}, \tag{4.18}$$

which now contains $\mathbf{u}$ explicitly. Therefore, Eq. (4.1) adopts the form of Eq. (4.18) in this case, with $\mathbf{z} = (\tilde{\mathbf{z}}, \tilde{\mathbf{u}})$, and

$$\boldsymbol{\Phi}(\mathbf{z}, \mathbf{u}) = \begin{bmatrix} \boldsymbol{\Psi}(\tilde{\mathbf{z}}, \tilde{\mathbf{u}}) \\ \tilde{\mathbf{u}} - \boldsymbol{\mu}(\mathbf{u}) \end{bmatrix}, \tag{4.19}$$

so that $\mathcal{W}$ will be the set of points $\mathbf{q} = (\mathbf{z}, \mathbf{u})$ satisfying Eq. (4.18) for which $\boldsymbol{\Phi}_{\mathbf{z}}$ is rank deficient.

Note now that, because of the form of Eq. (4.19), $\boldsymbol{\Phi}_{\mathbf{z}}$ has the block structure

$$\boldsymbol{\Phi}_{\mathbf{z}} = \left[\begin{array}{c|c} \boldsymbol{\Psi}_{\tilde{\mathbf{z}}} & \boldsymbol{\Psi}_{\tilde{\mathbf{u}}} \\ \hline \mathbf{0} & \mathbf{I}_{n_{\tilde{u}} \times n_{\tilde{u}}} \end{array}\right]$$

in this case, and therefore $\boldsymbol{\Phi}_{\mathbf{z}}$ is rank deficient if, and only if, its upper-left block $\boldsymbol{\Psi}_{\tilde{\mathbf{z}}}$ is rank deficient. In other words, $\mathcal{W}$ can be characterised as the set of points $\mathbf{q} = (\tilde{\mathbf{z}}, \tilde{\mathbf{u}}, \mathbf{u})$ that satisfy

$$\left.\begin{array}{r} \boldsymbol{\Psi}(\tilde{\mathbf{z}}, \tilde{\mathbf{u}}) = 0 \\ \tilde{\mathbf{u}} = \boldsymbol{\mu}(\mathbf{u}) \\ \boldsymbol{\Psi}_{\tilde{\mathbf{z}}}^{\mathsf{T}} \tilde{\boldsymbol{\xi}} = 0 \\ \tilde{\boldsymbol{\xi}}^{\mathsf{T}} \tilde{\boldsymbol{\xi}} = 1 \end{array}\right\} \tag{4.20}$$

for some $\tilde{\boldsymbol{\xi}}$, where $\tilde{\boldsymbol{\xi}}$ is a new vector with the suitable size.

While we could now proceed to isolate $\mathcal{W}$ by solving Eq. (4.20), $\boldsymbol{\mu}(\mathbf{u})$ usually introduces trigonometric terms that complicate the solution. Fortunately, since $\tilde{\mathbf{u}} = \boldsymbol{\mu}(\mathbf{u})$ parameterises the solution set of $\boldsymbol{\eta}(\tilde{\mathbf{u}}) = \mathbf{0}$, and the $\mathbf{u}$ variables only intervene in the second equation of (4.20), we can substitute $\tilde{\mathbf{u}} = \boldsymbol{\mu}(\mathbf{u})$ by $\boldsymbol{\eta}(\tilde{\mathbf{u}}) = \mathbf{0}$ in Eq. (4.20), arriving at the equivalent system

$$\left.\begin{array}{r} \boldsymbol{\Psi}(\tilde{\mathbf{z}}, \tilde{\mathbf{u}}) = \mathbf{0} \\ \boldsymbol{\eta}(\tilde{\mathbf{u}}) = \mathbf{0} \\ \boldsymbol{\Psi}_{\tilde{\mathbf{z}}}^{\mathsf{T}} \tilde{\boldsymbol{\xi}} = \mathbf{0} \\ \tilde{\boldsymbol{\xi}}^{\mathsf{T}} \tilde{\boldsymbol{\xi}} = 1 \end{array}\right\}. \tag{4.21}$$

Under the formulation of the assembly constraints explained in Chap. 3, both $\boldsymbol{\Psi}(\tilde{\mathbf{z}}, \tilde{\mathbf{u}})$ and $\boldsymbol{\eta}(\tilde{\mathbf{u}})$ are quadratic, meaning that Eq. (4.21) will be quadratic too. Hence, $\mathcal{W}$ can be characterised by Eq. (4.21) in this case.

4.4.3 Numerical Solution and Boundary Identification

As we have just seen, Eqs. (4.15) and (4.21) can both be written as a quadratic system of equations and, therefore, they can be solved using the same numerical method described in Sect. 3.3. To construct an initial box bounding their solution sets we only have to take into account that:

- The c_i variables can only take values in the range $[-1, 1]$, because they represent the cosine of some angle.
- The feasibility range of the t_i variables can be set to $[\,0, \sqrt{1-\cos\phi_i^l}\,]$ (Fig. 4.4, left).
- The range of the s_i variables can be set to $[0, h_i]$ (Fig. 4.4, right).
- Being unit vectors, the components of $\mathbf{k}_i$ and $\mathbf{n}_i$ can be bound to $[-1, 1]$.

As a result, the method will deliver a collection B of boxes tightly enclosing the solution set of the system. Using B, it is straightforward to obtain a box approximation $B^{\mathcal{W}}$ of the singular set $\mathcal{W}$. If B is obtained by solving Eq. (4.15), then $\mathbf{z}$ and $\mathbf{u}$ explicitly intervene as part of the variables in the system, and each box in B already has ranges along the $\mathbf{q} = (\mathbf{z}, \mathbf{u})$ dimensions. Such ranges define a box in the $\mathbf{q}$ space enclosing points of $\mathcal{W}$, and the collection of all such boxes directly provides $B^{\mathcal{W}}$. If B is instead obtained from Eq. (4.21), the $\mathbf{u}$ variables do not intervene in the variables of the system. However, for each box in B we can consider the ranges along the $\tilde{\mathbf{u}}$ variables, and derive corresponding ranges for the $\mathbf{u}$ variables by solving $\tilde{\mathbf{u}} = \boldsymbol{\mu}(\mathbf{u})$ using interval or linear relaxation techniques [28, 29].

Once $B^{\mathcal{W}}$ has been obtained, it remains to check whether the points of $\mathcal{W}$ enclosed in $B^{\mathcal{W}}$ correspond to boundary barriers, interior barriers, or traversable singularities. This classification is performed in two stages, illustrated in Fig. 4.6.

In a first stage, we classify the boxes of $B^{\mathcal{W}}$ according to whether they enclose barrier or traversable singularities. For each box $\mathcal{B}_i \in B^{\mathcal{W}}$ (Fig. 4.6a), a point $\mathbf{q}_i \in \mathcal{W}$ is computed (Fig. 4.6b), and the barrier determination method recalled in Sect. 4.2.2 is applied to this point. The computation of $\mathbf{q}_i$ is done by solving Eq. (4.15) or Eq. (4.21), depending on the situation, using a Newton-Raphson method starting from a point $\mathbf{q}_i^0$ taken from an arbitrary point $\mathbf{x}_i^0$ inside $\mathcal{B}_i$. This procedure will quadratically converge to some point $\mathbf{q}_i \in \mathcal{W}$ provided that the points within $\mathcal{B}_i$ are close enough to $\mathcal{W}$, which can be guaranteed by computing $B^{\mathcal{W}}$ using a sufficiently-small σ threshold (Sect. 3.3.3). The singularity type obtained for $\mathbf{q}_i$ (either barrier or traversable singularity) is taken as an estimation of the singularity type of all points in $\mathcal{B}_i \cap \mathcal{W}$, so that, after repeating this process for all boxes in $B^{\mathcal{W}}$, it is possible to subdivide $\mathcal{W}$ into subsets of constant singularity type. If $\mathbf{q}_i$ is a barrier singularity, a normal vector pointing towards the forbidden side of the barrier is drawn at $\mathbf{u}_i = \pi_{\mathbf{u}}(\mathbf{q}_i)$ (Fig. 4.6c).

In a second stage, we determine which of the barrier points $\mathbf{q}_i$ computed in the previous stage correspond to boundary or interior barriers. To do so, notice that $\pi_{\mathbf{u}}(\mathcal{W})$ subdivides $\mathcal{U}$ into several regions $\mathcal{R}_1, \ldots, \mathcal{R}_{n_r}$, where each region fully lies in the interior or in the exterior of $\mathcal{A}$, and a barrier point $\mathbf{u}_i$ will lie on $\partial\mathcal{A}$ if, and only if, one of its two neighbouring regions is exterior to $\mathcal{A}$. Thus, determining which of

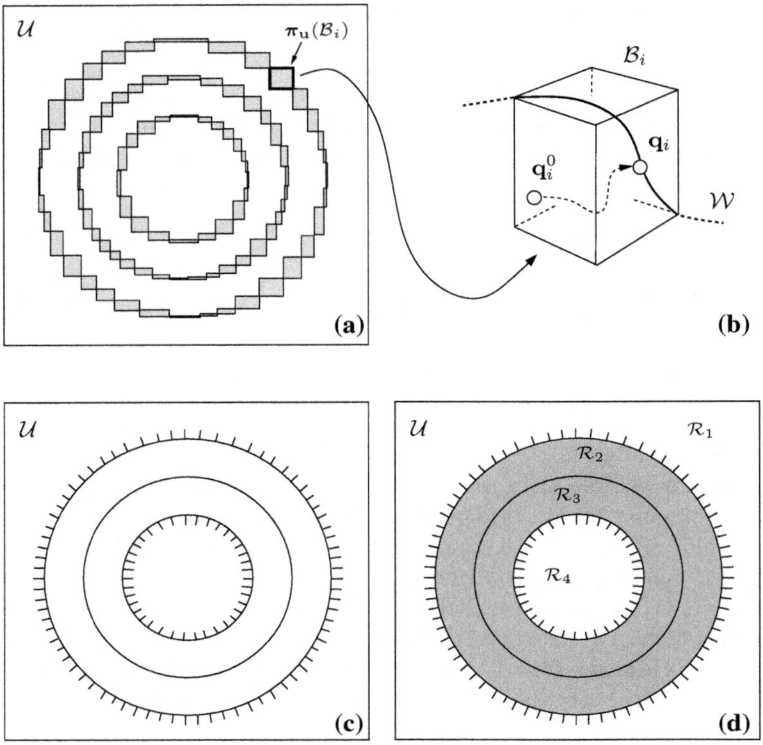

Fig. 4.6 Boundary identification process. (**a**) Box approximation of $\mathcal{W}$ projected to $\mathcal{U}$. (**b**) Computation of $\mathbf{q}_i \in \mathcal{W}$ for each box $\mathcal{B}_i$. (**c**) Classification of the points of $\pi_{\mathbf{u}}(\mathcal{W})$ into barrier or traversable singularities. (**d**) Regions into which $\pi_{\mathbf{u}}(\mathcal{W})$ subdivides $\mathcal{U}$ and their classification into interior (*grey*) or exterior (*white*) regions

the barrier points $\mathbf{u}_i$ correspond to boundary barriers boils down to checking whether the regions $\mathcal{R}_1, \ldots, \mathcal{R}_{n_r}$ are interior or exterior to $\mathcal{A}$. The type of a region $\mathcal{R}_j$ can then be determined by selecting a point $\mathbf{u}_j$ in the region, and solving $\mathbf{\Phi}(\mathbf{z}, \mathbf{u}_j) = \mathbf{0}$, which is here done with the same numerical technique proposed in Chap. 3. If $\mathbf{\Phi}(\mathbf{z}, \mathbf{u}_j) = \mathbf{0}$ has at least one solution, then $\mathcal{R}_j$ is an interior region, otherwise it is exterior.

While solving $\mathbf{\Phi}(\mathbf{z}, \mathbf{u}_j) = \mathbf{0}$ can be costly, we observe that it is not necessary to apply this test to most regions because the type of a region can often be decided with the following rules:

- If $\mathbf{u}$ only contains position coordinates of the end-effector, then the outer region will necessarily be exterior to $\mathcal{A}$, because the end-effector can only reach a bounded set of positions in practice.
- A region $\mathcal{R}_j$ whose boundary contains a traversable singularity can be marked as interior, because $\mathcal{R}_j$ contains trajectories that enter $\mathcal{R}_j$ through that singularity.

- A region $\mathcal{R}_j$ whose boundary contains a barrier point $\mathbf{u}_i$ with its normal vector pointing outwards from $\mathcal{R}_j$ can be marked as interior as well, because such barrier indicates that there are feasible trajectories in $\mathcal{C}$ projecting inside $\mathcal{R}_j$.

In Fig. 4.6d, for example, these observations allow us to identify $\mathcal{R}_1$ as an exterior region if $\mathbf{u}$ only contains position coordinates, and $\mathcal{R}_2$ and $\mathcal{R}_3$ as interior regions. Only the type of $\mathcal{R}_4$ needs to be disambiguated by checking a point in the region.

4.5 Case Studies

To emphasise the generality of the approach, and to encounter cases of multicomponent workspaces, hidden barriers, and degenerate barriers like those described in Sect. 4.3, we next compute several representative workspaces of planar and spatial mechanisms. As in Chap. 3, all computations have been done using the parallelized version of the method, implemented in C using the libraries of the CUIK Suite [30]. The method has been executed on a grid computer of Xeon processors able to run 160 threads in parallel. Table 4.1 summarises the size of the solved systems and the main performance data relative to the reported problems.

4.5.1 Multicomponent Workspaces

To give one example of a multicomponent workspace, we apply our method to obtain the reachable workspace of a planar mobility-three double-butterfly mechanism (Fig. 4.7). A version of this mechanism has been used to compare the performance of general position analysis methods [31–33], but no specific method has been given yet to compute the boundaries and interior barriers of its reachable workspace.

For this example we shall assume that the end-effector is the upper left body in Fig. 4.7, whose pose is determined by the position vector of point P and by the 2×2 rotation matrix of angle θ_1. We shall also suppose that two slider joints are mounted to allow the variation of lengths l_5 and l_7 within the intervals [11, 13] and [10, 12] respectively. As for the parameters in Fig. 4.7, we adopt the same values considered in [31–33]. Namely,

$$\begin{array}{lllll} a_0 = 7, & a_1 = 7, & a_2 = 5, & b_0 = 13, & b_1 = 6, \\ b_2 = 3, & l_3 = 7, & l_4 = 9, & a_6 = 3, & b_6 = 2, \\ \gamma_0 = 36.87°, & \gamma_1 = 22.62°, & \gamma_2 = 53.13°, & \gamma_6 = 36.87°, & \end{array}$$

where γ_i is the acute angle between segments a_i and b_i. For this mechanism, Eq. (4.1) can be formulated using the following equations:

1. The loop equations enforcing the closure of the three loops that leave the ground link via l_7, and return via l_4, l_3 and l_5 [32]:

$$
\begin{aligned}
0 &= l_7 c_7 + b_2 c_2 c_{\gamma_2} - b_2 s_2 s_{\gamma_2} - l_4 c_4 - a_6 c_6 + a_0 c_{\gamma_0},\\
0 &= l_7 s_7 + b_2 s_2 c_{\gamma_2} + b_2 c_2 s_{\gamma_2} - l_4 s_4 - a_6 s_6 - a_0 s_{\gamma_0},\\
0 &= l_7 c_7 + a_2 c_2 + a_1 c_1 - l_5 c_5 + b_0,\\
0 &= l_7 s_7 + a_2 s_2 + a_1 s_1 - l_5 s_5,\\
0 &= l_7 c_7 + a_2 c_2 + b_1 c_1 c_{\gamma_1} - b_1 s_1 s_{\gamma_1} - l_3 c_3 - b_6 c_6 c_{\gamma_6} + b_6 s_6 s_{\gamma_6} + a_0 c_{\gamma_0},\\
0 &= l_7 s_7 + a_2 s_2 + b_1 s_1 c_{\gamma_1} + b_1 c_1 s_{\gamma_1} - l_3 s_3 - b_6 s_6 c_{\gamma_6} - b_6 c_6 s_{\gamma_6} - a_0 s_{\gamma_0},
\end{aligned}
$$

   where c_{γ_i} and s_{γ_i} stand for the cosine and sine of γ_i, and c_i and s_i for those of θ_i.
2. The equations providing the x and y coordinates of P relative to the fixed OXY frame:

$$
\begin{aligned}
x &= b_0 + l_7 c_7 + a_2 c_2,\\
y &= l_7 s_7 + a_2 s_2.
\end{aligned}
$$

3. The circle equations constraining c_i and s_i:

$$c_i^2 + s_i^2 = 1.$$

4. The joint limit constraints for l_5 and l_7

$$(l_i - m_i)^2 + t_i^2 = h_i^2,$$

   where m_i and h_i are the mid-point and half-range of the intervals for l_5 and l_7.

The reachable workspace is defined as the set of attainable locations for a point on the end-effector, e.g. P in our case. Thus $\mathbf{u} = (x, y)$ for this workspace, and since x and y are explicit in the previous equations, we are in the situation of Eq. (4.15). Moreover, since $n_u = 2$, the boundary of the reachable workspace will be one-dimensional in the generic case.

The proposed method computes the box approximation of $\pi_{\mathbf{u}}(\mathcal{W})$ shown in Fig. 4.8, which delimits three workspace areas corresponding to different ways of assembling the mechanism. The result of the boundary identification process applied to one of such areas is shown in Fig. 4.9. Note that, having several connected components, this workspace would be difficult to map out entirely using the method in [23].

4.5.2 Hidden Barriers

To encounter workspaces with hidden barriers we next apply the method to the Gough-Stewart platform (Fig. 3.11). This is a challenging test case for any workspace

Table 4.1 Performance data relative to the reported examples

Figures	Mechanism		Workspace	Slice	d	N_{eq}–N_{var}	σ	T	N_{box}
4.8	Double butterfly		Reachable	–	1	29–30	0.1	255	14611
4.10	Gough-Stewart		Fixed orientation	–	2	17–19	0.01	236	558535
				$z = 4.95$	1	17–18	0.01	1	2337
				$z = 5$	1	17–18	0.01	1	2312
				$z = 5.10$	1	17–18	0.01	1	2260
				$z = 5.12$	1	17–18	0.01	2	2275
				$z = 5.145$	1	17–18	0.01	1	2305
				$z = 5.30$	1	17–18	0.01	1	2335
4.12 4.13	3-UPS/S		Orientation	–	2	24–26	0.03	1020	1021804
				$\sigma = -30°$	1	23–24	0.1	2	700
							0.01	8	5184
4.15 4.16	Agile Eye	$k = 0$	Orientation	–	1	24–26	0.01	1133	159246
		$k = \pi/36$		–	2	24–26	0.05	508	111910
		$k = \pi/12$		–	2	24–26	0.05	563	177522
		$k = \pi/6$		–	2	24–26	0.05	602	232461
4.18 4.19	15-link		Position	–	1	47–48	0.1	647	10312
				–	1	47–48	0.1	627	9810

For each example we indicate whether it entails computing the boundary and interior barriers of a whole workspace or just those of a particular slice. We also indicate the dimension of the set $\mathcal{W}$ computed (d), the number of equations (N_{eq}) and variables (N_{var}) defining this set, the accuracy threshold assumed (σ), the time employed in computing the set in seconds (T), and the number of solution boxes returned by the method (N_{box})

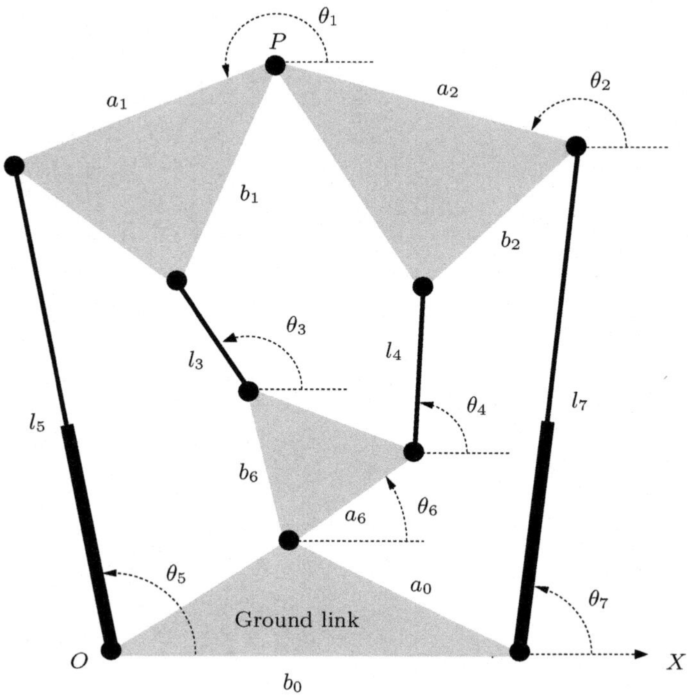

Fig. 4.7 The planar double-butterfly mechanism with variable lengths l_5 and l_7. The fixed frame is centred at O with the X axis aligned as indicated, and all angles θ_i are measured relative to such axis

determination method. The full workspace of this platform is six-dimensional, and its boundary five-dimensional, which hinders any attempt of computing it exhaustively. For this reason, and because six-dimensional sets are impossible to visualise directly in three dimensions, comprehension of this workspace can be gained by obtaining lower-dimensional subsets like

1. The *constant orientation* workspace, or set of attainable locations by a point P on the platform for a fixed platform orientation [6, 34].
2. The *constant position* workspace, or set of platform orientations for a fixed position of point P [10, 35–37].
3. The *reachable* workspace, or set of locations that P can attain with at least one platform orientation [38, 39].

All of these workspaces can be computed by the proposed technique using a proper choice of the **u** variables and fixing others to given values. As an example, we next compute a constant orientation workspace studied in [6].

Equation (4.1) can be formulated as done in Sect. 3.5.2. Here, the rotation matrix of the platform, $\mathbf{R}_2$, will be constant and known, since we are computing the con-

Fig. 4.8 Box approximation of the set $\pi_{\mathbf{u}}(\mathcal{W})$ corresponding to the reachable workspace of the mechanism in Fig. 4.7

stant orientation workspace. Also, since the lengths d_i will be limited within ranges $\left[\,\underline{d_i}, \overline{d_i}\,\right]$, we must add the equations

$$(d_i - m_i)^2 + t_i^2 = h_i^2 \tag{4.22}$$

to the system, for $i = 1, \ldots, 6$, where m_i and h_i are the mid-point and half-range of $\left[\,\underline{d_i}, \overline{d_i}\,\right]$. If the position vector of the reference point P is $\mathbf{r}_P = (x, y, z)$, we have $\mathbf{u} = (x, y, z)$, and we are in the situation in which $\mathbf{u}$ intervenes explicitly in $\mathbf{q}$. Thus, the system characterising $\mathcal{W}$ adopts the form of Eq. (4.15).

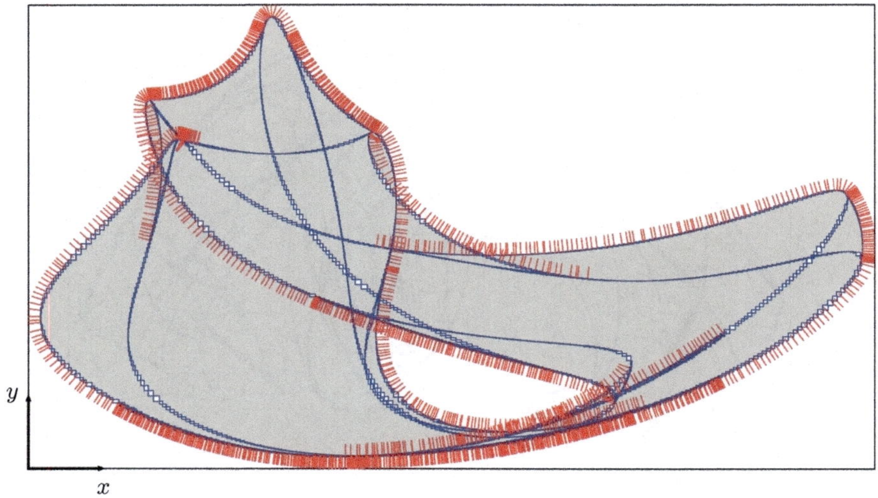

Fig. 4.9 Results of applying the barrier analysis process from Sect. 4.4.3 to one of the curve components in Fig. 4.8. The same conventions as in Figs. 4.2 and 4.6d are used

Figure 4.10 shows 3-D views of the box approximation obtained for $\boldsymbol{\pi}_{\mathbf{u}}(\mathcal{W})$, which describes an umbrella-like surface overall. The computation was done assuming the parameters in Table 4.2, with $\mathbf{R}_2$ fixed to the identity matrix, and $[\underline{d_i}, \overline{d_i}] = [454.5, 504.5]$. In fact, this workspace has an additional connected component symmetric to the one of Fig. 4.10, which corresponds to the assembly of this mechanism in which P sweeps a similar volume for $z < 0$. All results obtained are consistent with those in [6].

To better appreciate the shape of the enclosed volume, Fig. 4.10 plots constant-z slices of $\boldsymbol{\pi}_{\mathbf{u}}(\mathcal{W})$ indicating the results of the boundary identification process. Notice from the plots that it would be difficult to compute such slices by continuation [23],

Table 4.2 Parameters of the Gough-Stewart platform studied in [6]

Leg	Base L_1	Platform L_2
1	$\mathbf{p}_{1,1}^{\mathcal{F}_1} = (92.58, 99.64, 23.10)$	$\mathbf{p}_{2,1}^{\mathcal{F}_2} = (30.00, 73.00 - 37.10)$
2	$\mathbf{p}_{1,2}^{\mathcal{F}_1} = (132.58, 30.36, 23.10)$	$\mathbf{p}_{2,2}^{\mathcal{F}_2} = (78.22, -10.52, -37.10)$
3	$\mathbf{p}_{1,3}^{\mathcal{F}_1} = (40.00, -130.00, 23.10)$	$\mathbf{p}_{2,3}^{\mathcal{F}_2} = (48.22, -62.48, -37.10)$
4	$\mathbf{p}_{1,4}^{\mathcal{F}_1} = (-40.00, -130.00, 23.10)$	$\mathbf{p}_{2,4}^{\mathcal{F}_2} = (-48.22, -62.48, -37.10)$
5	$\mathbf{p}_{1,5}^{\mathcal{F}_1} = (-132.58, 30.36, 23.10)$	$\mathbf{p}_{2,5}^{\mathcal{F}_2} = (-78.22, -10.52, -37.10)$
6	$\mathbf{p}_{1,6}^{\mathcal{F}_1} = (-92.58, 99.64, 23.10)$	$\mathbf{p}_{2,6}^{\mathcal{F}_2} = (-30.00, 73.00, -37.10)$

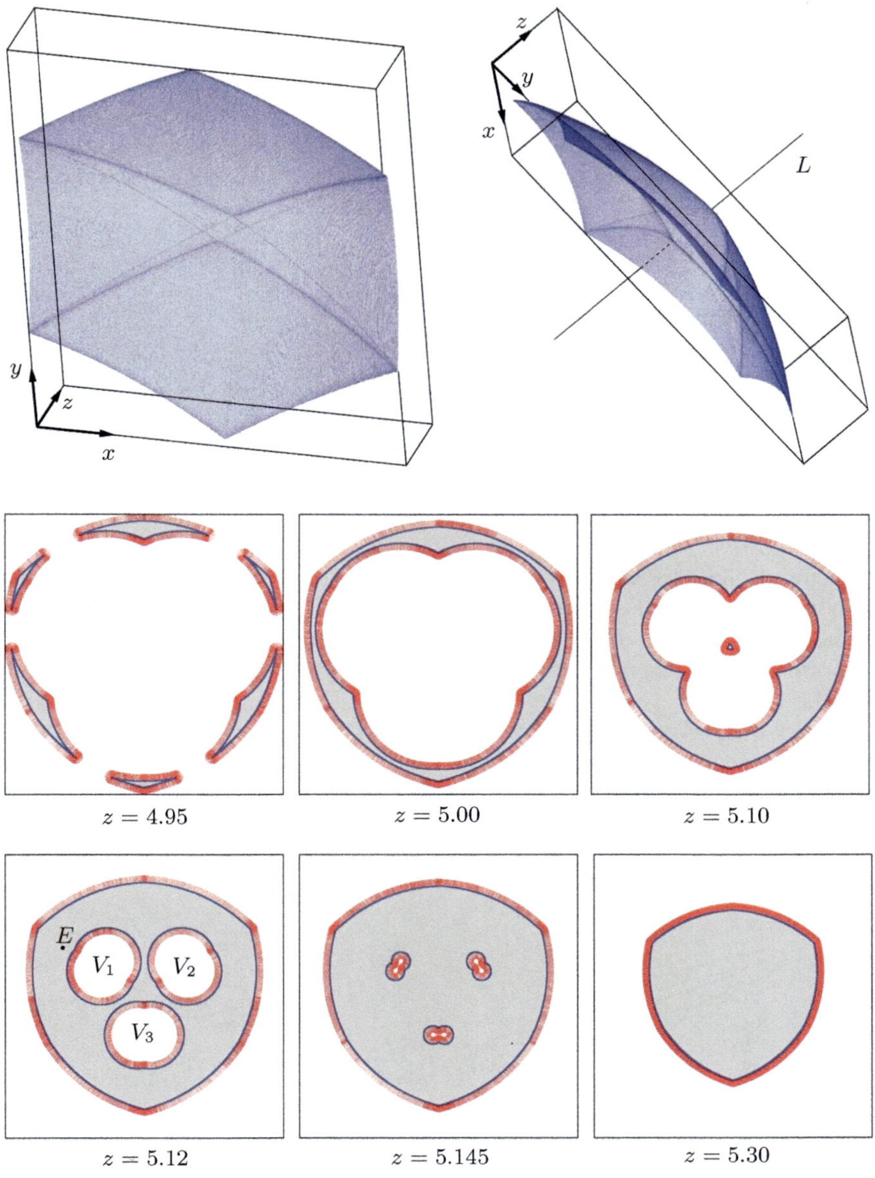

Fig. 4.10 (*Top row*) Two views of the boundary of a constant orientation workspace of a Gough-Stewart platform. *Boxes* are translucent to better appreciate the shape. (*Remaining rows*) Slices of the workspace for different values of z. All points of $\mathcal{W}$ are classified as barrier singularities in this case

because many of the slices present multicomponent boundaries and hidden barriers that hinder the application of the ray-shooting technique described in Sect. 4.3. If, for example, the ray is shot from point E on the $z = 5.12$ slice, it will not hit the boundary of voids V_2 and V_3. While it is true that in [34] the authors were able to compute $\pi_{\mathbf{u}}(\mathcal{W})$ using the method in [23], they did so by defining particular slices of this set obtained by cutting the umbrella with planes through line L shown in Fig. 4.10. This solution avoids the appearance of internal voids within each slice, but obviously relies on a priori knowledge of the result.

4.5.3 Degenerate Barriers

Because of the complexity of their defining equations, orientation workspaces are considered among the most difficult ones to compute and represent [10, 36, 37, 40]. We shall illustrate their obtention by means of spherical parallel mechanisms because this will lead to one example of the degenerate barriers mentioned in Sect. 4.3, which the method in [23] is unable to identify. The examples are taken from [12] and correspond to well-known architectures of three degree-of-freedom spherical parallel mechanisms: the 3-UPS/S and 3-RRR designs depicted in Fig. 4.11a and b. These are orientational mechanisms where the moving platform can be rotated with respect to the base about a fixed point O by actuating some of the leg joints. We next compute their orientation workspace and verify the results with those of the analytic method in [12].

To derive Eq. (4.1), note that, irrespectively of the chosen architecture, each leg imposes the same constraint on the moving platform. In a 3-UPS/S platform, for example, d_i is constrained to take values within some interval $\left[\,\underline{d_i}, \overline{d_i}\,\right]$ by design, which limits the angle between $OP_{1,i}$ and $OP_{2,i}$ to some range $\left[\,\underline{\alpha_i}, \overline{\alpha_i}\,\right]$. In a 3-RRR platform, the angle between $OP_{1,i}$ and $OP_{2,i}$ is also limited to a range $\left[\,\underline{\alpha_i}, \overline{\alpha_i}\,\right]$ due

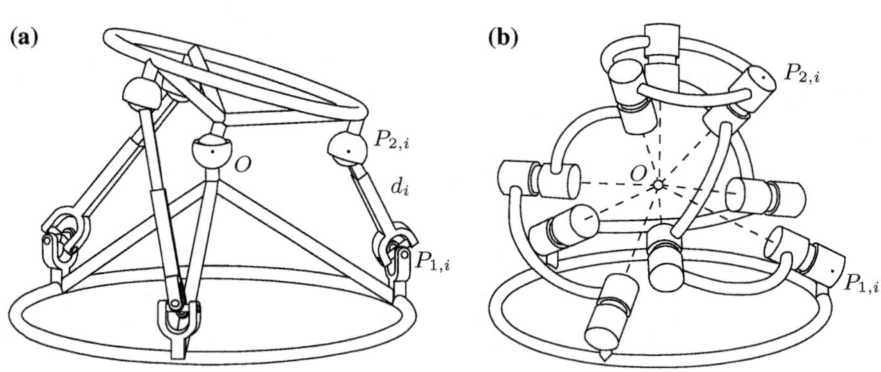

Fig. 4.11 The 3-UPS/S and 3-RRR spherical platforms. Figures adapted from [12]

to mechanical limits in the joints, or to the angles encompassed by the leg links. Both designs are thus kinematically equivalent. Moreover, the 3-UPS/S design can be obtained as a special case of the Gough-Stewart platform by making three anchor points of such platform coincident, and locking the corresponding legs. Hence, for both the 3-UPS/S and 3-RRR designs, Eq. (4.1) can be formulated as in the previous Section but assuming that $P = O$, and that both the fixed and moving reference frames are centred in O.

In general, the orientation workspace is defined as the set of possible values for three orientation angles of the platform. Although any set of Euler angles could be used, we shall here adopt the azimuth (ϕ), tilt (θ), and torsion (σ) angles assumed in [12] to ease the comparison of results. Using such angles,

$$\mathbf{R}_2 = \mathbf{R}_z(\phi)\mathbf{R}_y(\theta)\mathbf{R}_z(\gamma),$$

where $\gamma = \sigma - \phi$, and thus the columns of $\mathbf{R}_2$ are

$$\mathbf{r}_1 = \begin{bmatrix} \cos\phi\cos\theta\cos\gamma - \sin\phi\sin\gamma \\ \sin\phi\cos\theta\cos\gamma + \cos\phi\sin\gamma \\ -\sin\theta\cos\gamma \end{bmatrix}, \tag{4.23}$$

$$\mathbf{r}_2 = \begin{bmatrix} -\cos\phi\cos\theta\sin\gamma - \sin\phi\cos\gamma \\ -\sin\phi\cos\theta\sin\gamma + \cos\phi\cos\gamma \\ \sin\theta\sin\gamma \end{bmatrix}, \tag{4.24}$$

$$\mathbf{r}_3 = \begin{bmatrix} \cos\phi\sin\theta \\ \sin\phi\sin\theta \\ \cos\theta \end{bmatrix}. \tag{4.25}$$

Under the previous convention, Bonev and Gosselin define the orientation workspace as the set of possible values that

$$\mathbf{u} = (\phi, \theta, \sigma)$$

can attain [12]. We are thus in the situation in which $\mathbf{u}$ does not intervene explicitly in $\mathbf{q}$, but it can be related to

$$\tilde{\mathbf{u}} = (\mathbf{r}_1, \mathbf{r}_2, \mathbf{r}_3)$$

using Eqs. (4.23)–(4.25). Therefore, $\boldsymbol{\eta}(\tilde{\mathbf{u}}) = \mathbf{0}$ is given by Eqs. (3.8)–(3.11) in this case, and $\boldsymbol{\Psi}(\tilde{\mathbf{z}}, \tilde{\mathbf{u}}) = \mathbf{0}$ consists of Eqs. (3.31), (3.32) and (4.22), with

$$\tilde{\mathbf{z}} = (\mathbf{d}_1, \mathbf{d}_2, \mathbf{d}_3, d_1, d_2, d_3, t_1, t_2, t_3).$$

Overall, Eq. (4.21) contains 37 equations in 39 variables, and the boundaries of the workspace are hence expected to be two-dimensional.

For the particular examples shown next we assume the same symmetry conditions as in [12]. Namely, that $P_{1,i}$ and $P_{2,i}$ will lie on a unit sphere centred at O, with position vectors

$$\begin{bmatrix} \cos((i-1)\frac{2\pi}{3})\sin\beta_1 \\ \sin((i-1)\frac{2\pi}{3})\sin\beta_1 \\ -\cos\beta_1 \end{bmatrix}, \quad \begin{bmatrix} \cos((i-1)\frac{2\pi}{3})\sin\beta_2 \\ \sin((i-1)\frac{2\pi}{3})\sin\beta_2 \\ -\cos\beta_2 \end{bmatrix},$$

respectively, and we shall set $\left[\,\underline{\alpha_i}, \overline{\alpha_i}\,\right] = \left[\,\underline{\alpha}, \overline{\alpha}\,\right]$ for all i.

Figure 4.12 shows the resulting box approximation of $\boldsymbol{\pi}_{\mathbf{u}}(\mathcal{W})$ for the values

$$\beta_1 = 45^\circ, \qquad \beta_2 = 35^\circ, \qquad \underline{\alpha} = 20^\circ, \qquad \overline{\alpha} = 130^\circ.$$

These parameters correspond to one of the cases analysed in [12], where constant-torsion slices of $\boldsymbol{\pi}_{\mathbf{u}}(\mathcal{W})$ are provided for these mechanisms. As expected, $\boldsymbol{\pi}_{\mathbf{u}}(\mathcal{W})$ is a surface in the (θ, ϕ, σ) space and, by using the neighbouring relationships of the returned boxes, this surface can be shown to contain just one connected component. Figure 4.13 shows, on the left, the $\sigma = -30^\circ$ slice of the surface shown in Fig. 4.12, and the barriers identified on such slice. The resulting curve and the interior regions detected match those in [12] when plotted in polar coordinates (Fig. 4.13, right).

While the orientation workspace will generally have a two-dimensional boundary, such boundary may degenerate into one-dimensional barriers for particular choices of the geometric parameters, which poses serious difficulties to the continuation method in [23]. This is what occurs in the Agile Eye mechanism shown in Fig. 4.14 for example [42, 43], a well-known instance of the 3-RRR design in Fig. 4.11b, in which

$$\beta_1 = \beta_2 = \arccos(1/\sqrt{3}), \qquad \underline{\alpha} = 0, \qquad \overline{\alpha} = \pi.$$

As mentioned in Sect. 4.3, computing such barriers using [23] is almost impossible in this mechanism because the ray shooting technique will fail to converge to the barriers almost always. On the contrary, the presented technique is immune to such situations. If the same equations considered for obtaining the plot in Fig. 4.12 are now used for determining the workspace boundaries of the Agile Eye relative to $\mathbf{u} = (\phi, \theta, \sigma)$, we readily obtain the curves depicted in Fig. 4.15, which agree with those described in [12]. When analysed, these curves are seen to be barrier singularities interior to the workspace. In other words, the mechanism will be able to reach any possible orientation, but it will find a motion impediment when trying to traverse across the curves. The phenomenon is better understood if we view the curves as the limit case of a one-parameter family of workspace boundaries, corresponding to setting

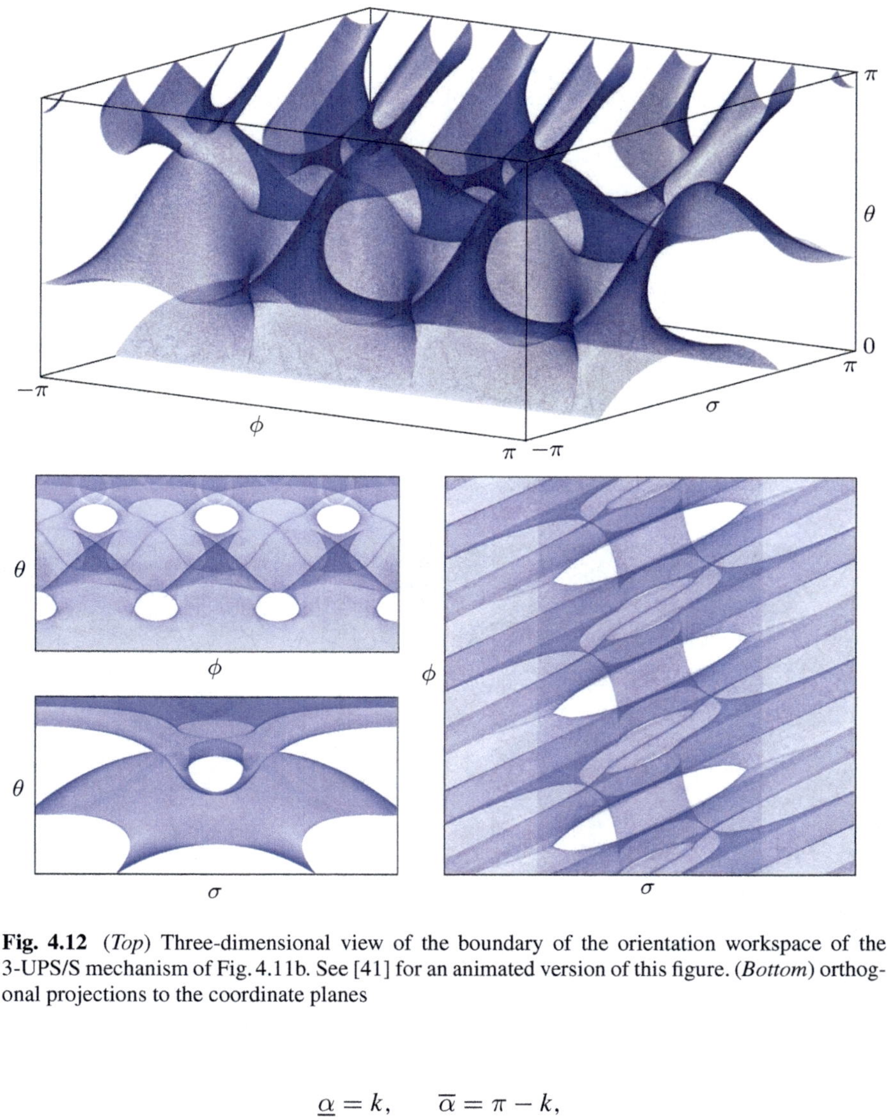

Fig. 4.12 (*Top*) Three-dimensional view of the boundary of the orientation workspace of the 3-UPS/S mechanism of Fig. 4.11b. See [41] for an animated version of this figure. (*Bottom*) orthogonal projections to the coordinate planes

$$\underline{\alpha} = k, \qquad \overline{\alpha} = \pi - k,$$

and varying the parameter k from $\frac{\pi}{6}$ to 0, while keeping fixed the remaining set of geometric parameters. Tubular boundary barriers are present in all members of the family (first two plots of Fig. 4.16), which degenerate into the barrier filaments of the Agile Eye (last plot).

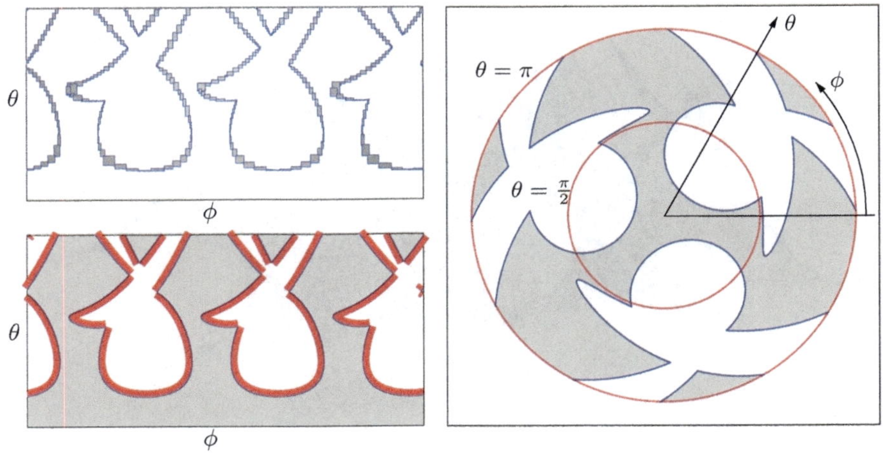

Fig. 4.13 Constant torsion workspace of an instance of the spherical mechanism of Fig. 4.11b

Fig. 4.14 The Agile Eye mechanism for the rapid orientation of a camera. Photo courtesy of Prof. Clément Gosselin

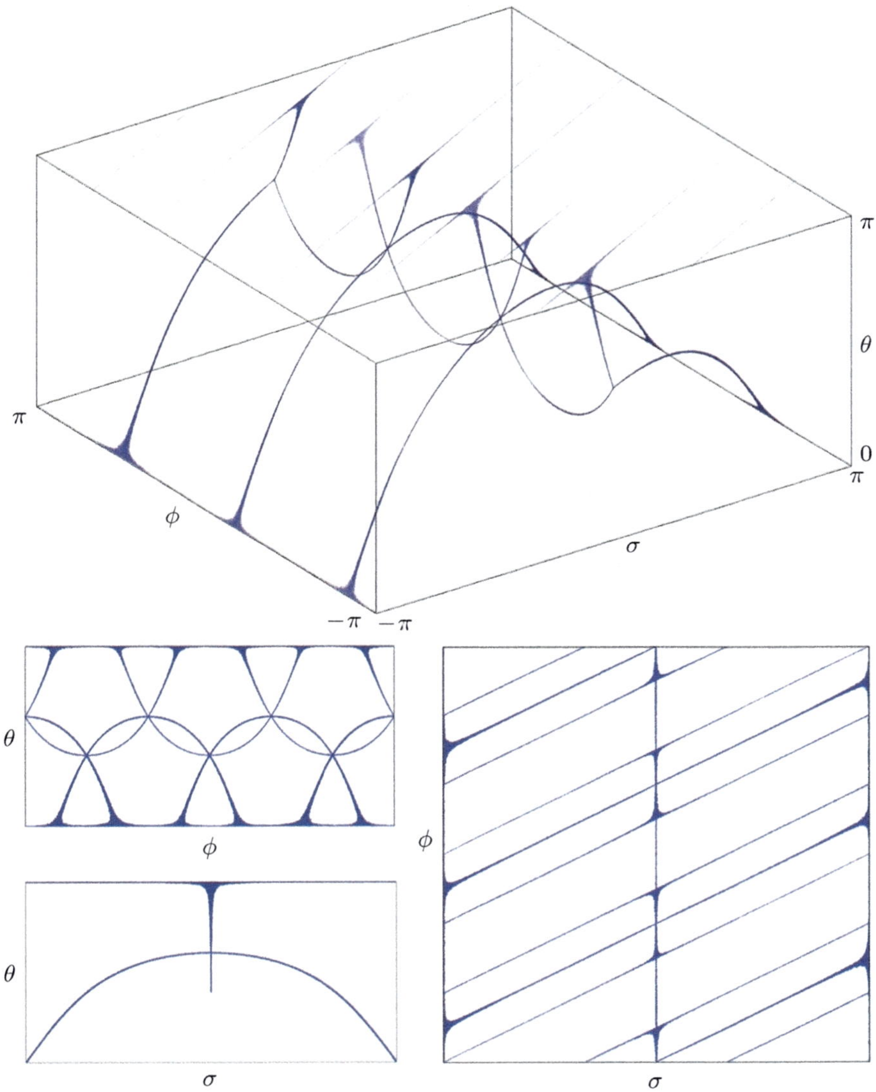

Fig. 4.15 The workspace of the Agile Eye has degenerate barriers. Each line in the $\theta = \pi$ plane corresponds to a unique pose of the platform. See [41] for a video showing these barriers from different viewpoints

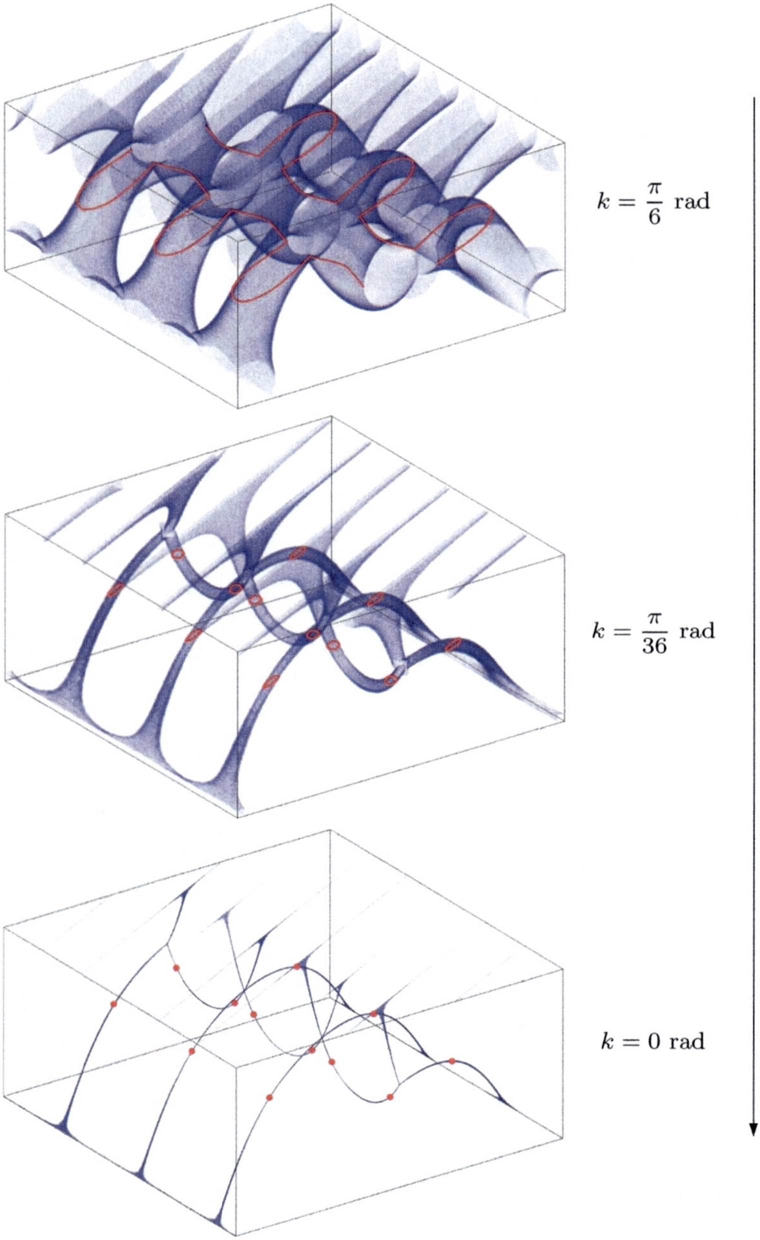

Fig. 4.16 The workspace of the Agile Eye is a limit case of a one-parameter family of workpaces. A constant-θ slice is shown in *red* in each plot, to see how it evolves from a one-dimensional curve (*top plot*) to isolated points (*bottom plot*)

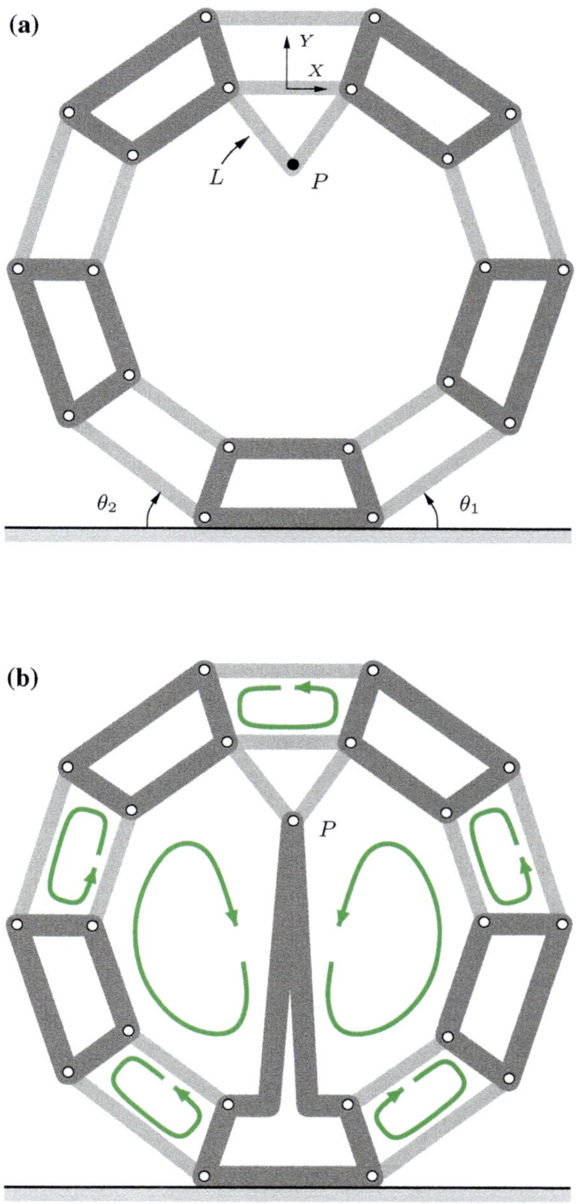

Fig. 4.17 (**a**) A 15-link mechanism. (**b**) Its inverse kinematic problem is equivalent to solving the position analysis of a seven-loop truss

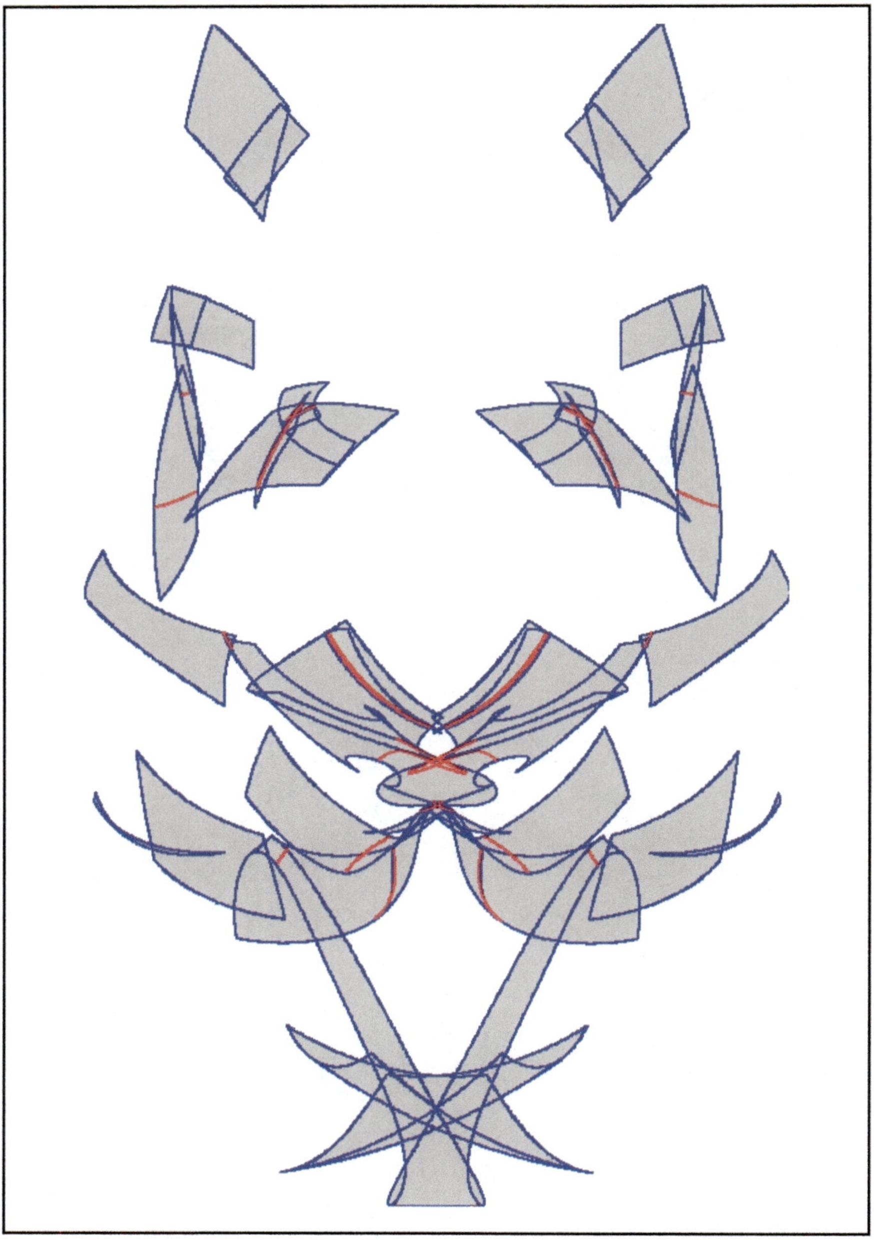

Fig. 4.18 Position workspace of the mechanism in Fig. 4.17, using the geometric parameters mentioned in the text, and assuming that θ_1 and θ_2 are limited to the range $[45.57°, 60°]$. The *red curves* correspond to the forward singularities of the mechanism, while the inverse singularities are included among the *blue curves*

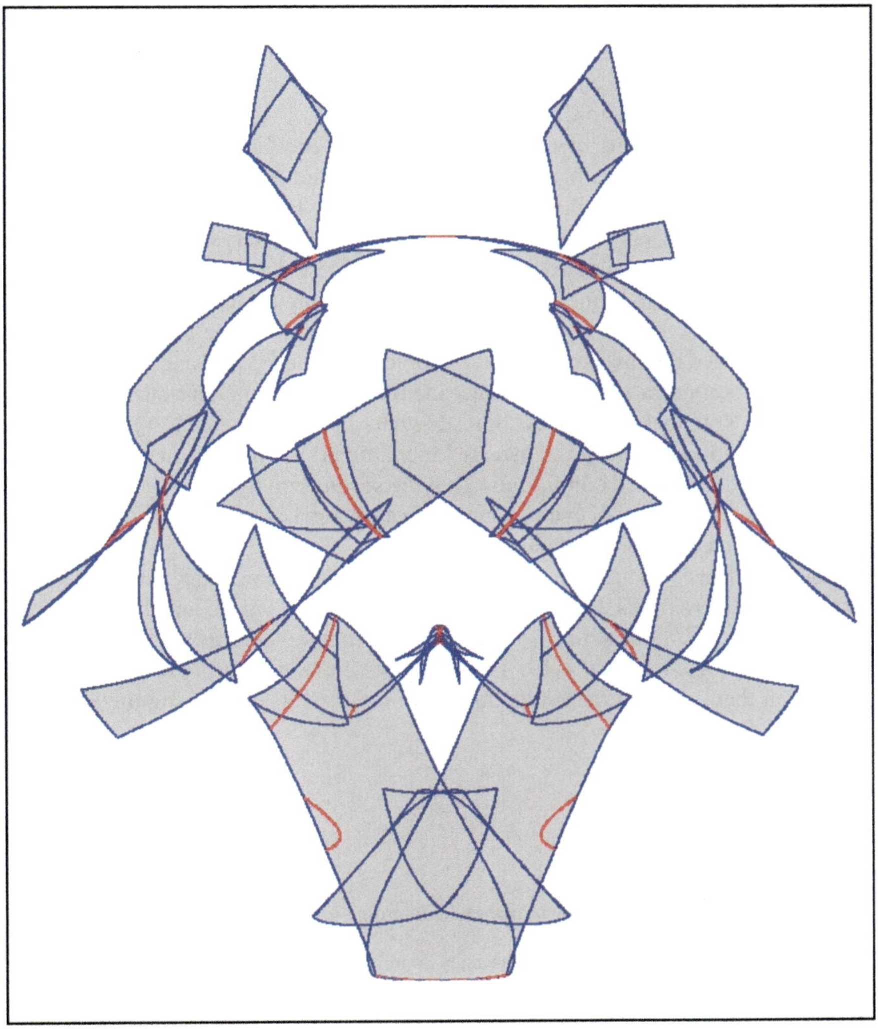

Fig. 4.19 Position workspace of the mechanism in Fig. 4.17, but now with the angles θ_1 and θ_2 limited to the range [36.87°, 53.13°]

4.5.4 Very Complex Mechanisms

To illustrate the method in a highly complex situation, we next apply it to compute the position workspace of a point of the 15-link mechanism in Fig. 4.17a. The mechanism consists of five quadrilateral links interconnected by means of bar links and revolute joints, forming a decagonal ring. If we fix one of the quadrilaterals to ground, the mechanism has mobility two, so that $\mathcal{C}$ will have dimension $n = 2$ in general, and $\mathcal{W}$ will be formed by one or several curves in such space. The mechanism is controlled by actuating the indicated angles θ_1 and θ_2, and the output of interest is the position of a point $P = (x, y)$ on link L, so that $\mathbf{u} = (x, y)$.

The complexity of this mechanism comes from the fact that it involves many links, and all of them move in a highly-coupled manner. This behaviour is apparent from the structure of the mechanism already, but it can be proved by applying recent Assur graph theory tools [44, 45]. On the basis of these observations, we conjecture that the derivation of minimal-degree polynomials describing the singularity sets of this mechanism is an extremely difficult task. The computation of such sets is even too hard using discretisation techniques [46, 47], which define a grid of points in the $\mathcal{U}$ space, solve the inverse kinematic problem for each point, and finally analyse the resulting configurations one-by-one, identifying those that are close to the singularity sets. Note that this process boils down to discretising the (x, y) plane in this mechanism, and that solving the inverse kinematic problem for each position (x, y) is equivalent to finding all configurations of the seven-loop truss in Fig. 4.17b, which is beyond the capabilities of even the most advanced techniques for position analysis based oncharacteristic polynomials [48, 49].

Assuming that P is located in position $(0, -1)$ of the frame indicated in Fig. 4.17a, that all quadrilateral links of the mechanism are squares of side 1, and that all bars are of length 2, except L, which is of length $\sqrt{2}$, the method determines the singularity sets shown in Figs. 4.18 and 4.19. The two plots correspond to two variants of the mechanism that differ on the limits imposed on θ_1 and θ_2 only, using the equations of Sect. 4.4.1.

References

1. B. Roth, Performance evaluation of manipulators from a kinematic viewpoint. Nat. Bur. Stand. Spec. Publ. **459**, 39–62 (1976)
2. F. Freudenstein, E.J.F. Primrose, On the analysis and synthesis of the workspace of a three-link, turning-pair connected robot arm. ASME J. Mech. Des. **106**(3), 365–370 (1984)
3. A.H. Soni, Y.C. Tsai, An algorithm for the workspace of a general n-R robot. ASME J. Mech. Des. **105**, 52–57 (1983)
4. J.A. Hansen, K.C. Gupta, S.M.K. Kazerounian, Generation and evaluation of the workspace of a manipulator. Int. J. Robot. Res. **2**(3), 22–31 (1983)
5. J. Rastegar, B. Fardanesh, Manipulation workspace analysis using the Monte Carlo method. Mech. Mach. Theor. **25**(2), 233–239 (1990)
6. C.M. Gosselin, Determination of the workspace of 6-DOF parallel manipulators. ASME J. Mech. Des. **112**, 331–336 (1990)
7. J.-P. Merlet, Geometrical determination of the workspace of a constrained parallel manipulator, in *Advances in Robot Kinematics* (1992), pp. 326–329
8. J.-P. Merlet, C.M. Gosselin, N. Mouly, Workspaces of planar parallel manipulators. Mech. Mach. Theor. **33**(1–2), 7–20 (1998)
9. J.-P. Merlet, Determination of 6D workspaces of Gough-type parallel manipulator and comparison between different geometries. Int. J. Robot. Res. **18**(9), 902–916 (1999)
10. I.A. Bonev, J.-C. Ryu, A new approach to orientation workspace analysis of 6-DOF parallel manipulators. Mech. Mach. Theor. **36**(1), 15–28 (2001)

11. J.A. Snyman, L.J. du Plessis, J. Duffy, An optimization approach to the determination of the boundaries of manipulator workspaces. ASME J. Mech. Des. **122**(4), 447–456 (2000)
12. I.A. Bonev, C.M. Gosselin, Analytical determination of the workspace of symmetrical spherical parallel mechanisms. IEEE Trans. Robot. **22**(5), 1011–1017 (2006)
13. K. Abdel-Malek, H.J. Yeh, Analytical boundary of the workspace for general 3-DOF mechanisms. Int. J. Robot. Res. **16**(2), 198–213 (1997)
14. M. Ceccarelli, A formulation for the workspace boundary of general N-revolute manipulators. Mech. Mach. Theor. **31**(5), 637–646 (1996)
15. K. Abdel-Malek, H.J. Yeh, N. Khairallah, Workspace, void, and volume determination of the general 5-DOF manipulator. J. Struct. Mech. **27**(1), 89–115 (1999)
16. M. Zein, P. Wenger, D. Chablat, An exhaustive study of the workspace topologies of all 3R orthogonal manipulators with geometric simplifications. Mech. Mach. Theor. **41**(8), 971–986 (2006)
17. E. Ottaviano, M. Husty, M. Ceccarelli, Identification of the workspace boundary of a general 3-R manipulator. ASME J. Mech. Des. **128**(1), 236–242 (2006)
18. D.-Y. Jo, *Numerical Analysis of Workspaces of Multibody Mechanical Systems*. Ph.D. Thesis (The University of Iowa, Iowa, 1988)
19. F. A. Adkins, *Numerical Continuation and Bifurcation Methods for Mechanism Workspace and Controllability Issues*. Ph.D. Thesis (The University of Iowa, Iowa, 1996)
20. K. Abdel-Malek, J. Yang, D. Blackmore, K. Joy, Swept volumes: fundation, perspectives, and applications. Int. J. Shape Model. **12**(1), 87–127 (2006)
21. Y. Lu, Y. Shi, B. Hu, Solving reachable workspace of some parallel manipulators by computer-aided design variation geometry. Proc. Inst. Mech. Eng. Part C: J. Mech. Eng. Sci. **222**(9), 1773–1781 (2008)
22. G. Castelli, E. Ottaviano, M. Ceccarelli, A fairly general algorithm to evaluate workspace characteristics of serial and parallel manipulators. Mech. Based Des. Struct. Mach. **36**(1), 14–33 (2008)
23. E.J. Haug, C.-M. Luh, F.A. Adkins, J.-Y. Wang, Numerical algorithms for mapping boundaries of manipulator workspaces. ASME J. Mech. Des. **118**(2), 228–234 (1996)
24. K. Abdel-Malek, F.A. Adkins, H.J. Yeh, E.J. Haug, On the determination of boundaries to manipulator workspaces. Robot. Comput. Integr. Manuf. **13**(1), 63–72 (1997)
25. D. Oblak, D. Kohli, Boundary surfaces, limit surfaces, crossable and noncrossable surfaces in workspace of mechanical manipulators. ASME J. Mech. Des. **110**(4), 389–396 (1988)
26. S.G. Krantz, H.R. Parks, *The Implicit Function Theorem: History* (Theory and Applications, Birkhäuser, 2002)
27. E.L. Allgower, K. Georg, *Numerical Continuation Methods* (Springer, 1990)
28. R.E. Moore, R.B. Kearfott, M.J. Cloud, *Introduction to Interval Analysis* (Society for Industrial Mathematics, 2009)
29. J.M. Porta, L. Ros, F. Thomas, A linear relaxation technique for the position analysis of multi-loop linkages. IEEE Trans. Robot. **25**(2), 225–239 (2009)
30. The CUIK Project Home Page, http://www.iri.upc.edu/cuik. Accessed 16 Jun 2016
31. C.W. Wampler, Solving the kinematics of planar mechanisms by Dixon's determinant and a complex plane formulation. ASME J. Mech. Des. **123**, 382–387 (2001). September
32. J.M. Porta, L. Ros, T. Creemers, F. Thomas, Box approximations of planar linkage configuration spaces. ASME J. Mech. Des. **129**(4), 397–405 (2007)
33. E. Celaya, T. Creemers, L. Ros, Exact interval propagation for the efficient solution of position analysis problems on planar linkages. Mech. Mach. Theor. **54**, 116–131 (2012)
34. D.Y. Jo, E.J. Haug, Workspace analysis of closed-loop mechanisms with unilateral constraints. Adv. Des. Autom. **19**(3), 53–60 (1989)
35. J.-P. Merlet, Determination of the orientation workspace of parallel manipulators. J. Intell. Robot. Syst. **13**(2), 143–160 (1995)
36. Q. Jiang, C.M. Gosselin, Determination of the maximal singularity-free orientation workspace for the Gough-Stewart platform. Mech. Mach. Theor. **44**(6), 1281–1293 (2009)

37. Y. Cao, Z. Huang, H. Zhou, W. Ji, Orientation workspace analysis of a special class of Stewart-Gough parallel manipulators. Robotica **28**(7), 989–1000 (2010)
38. C.-M. Luh, F.A. Adkins, E.J. Haug, C.C. Qiu, Working capability analysis of Stewart platforms. ASME J. Mech. Des. **118**(2), 220–227 (1996)
39. F. Pernkopf, M. Husty, Workspace analysis of Stewart-Gough-type parallel manipulators. Proc. Inst. Mech. Eng. Part C: J. Mech. Eng. Sci. **220**(7), 1019–1032 (2006)
40. J.-P. Merlet, C.M. Gosselin, *Springer Handbook of Robotics*, ch. Parallel Mechanisms and Robots (Springer, 2008), pp. 269–285
41. Companion web page of this book: http://www.iri.upc.edu/srm. Accessed 16 Jun 2016
42. C.M. Gosselin, E. St.-Pierre, Development and experimentation of a fast three-degree-of-freedom camera-orienting device. *International Journal of Robotics Research*, vol. 16, no. 5 (1997), pp. 619–630
43. I.A. Bonev, D. Chablat, P. Wenger, Working and assembly modes of the Agile Eye, in *Proceedings of the IEEE International Conference on Robotics and Automation, ICRA* (Orlando, USA, 2006), pp. 2317–2322
44. B. Servatius, O. Shai, W. Whiteley, Combinatorial characterization of the Assur graphs from engineering. Eur. J. Comb. **31**(4), 1091–1104 (2010)
45. A. Sljoka, O. Shai, W. Whiteley, Checking mobility and decomposition of linkages via pebble game algorithms, in *Proceedings of the ASME International Design Engineering Technical Conferences and Computers and Information in Engineering Conference, IDETC/CIE* (Washington, USA, 2011), pp. 493–502
46. O. Altuzarra, C. Pinto, R. Avilés, A. Hernández, A practical procedure to analyze singular configurations in closed kinematic chains. IEEE Trans. Robot. **20**(6), 929–940 (2004)
47. E. Macho, O. Altuzarra, E. Amezua, A. Hernández, Obtaining configuration space and singularity maps for parallel manipulators. Mech. Mach. Theor. **44**(11), 2110–2125 (2009)
48. N. Rojas, F. Thomas, On closed-form solutions to the position analysis of Baranov trusses. Mech. Mach. Theor. **50**, 179–196 (2011)
49. N. Rojas, F. Thomas, Distance-based position analysis of the three seven-link Assur kinematic chains. Mech. Mach. Theor. **46**(2), 112–126 (2010)

Chapter 5
Singularity-Free Path Planning

The methods in the previous chapters allow us to compute the different singularity sets of a mechanism. The visualisation of the sets is not an easy task, but we have shown how the set projections to the input and output spaces generate very rich diagrams, called portraits, summarising the global motion capabilities of a mechanism. These diagrams constitute safe navigation maps because any path in them avoiding a projected singularity always corresponds to a singularity-free path in the C-space. However, we have seen that they do not always reveal all possible singularity-free paths between two configurations. Additional tools to determine such paths become necessary for a complete understanding of the motion capabilities of a robot.

This chapter provides an algorithm for this purpose. Given two nonsingular configurations of the mechanism, the algorithm attempts to connect them through a path that maintains a minimum clearance with respect to the singularity locus at all points. In comparison to previously-existing approaches (Sect. 5.1), the method can be applied to nonredundant mechanisms of general architecture and it is resolution-complete—in the sense that it always returns a path whenever one exists at a given resolution. The strategy relies on defining a smooth manifold that maintains a one-to-one correspondence with the singularity-free C-space of the mechanism (Sect. 5.2), and on using a higher-dimensional continuation technique to explore this manifold systematically from one configuration, until the second configuration is found, or path nonexistence is determined by exhaustion of the search (Sect. 5.3). Following the aim of [1], the method can also be used to compute a full atlas of the singularity-free manifold that is reachable from one configuration, which is useful to rapidly resolve planning queries within such manifold, or to visualise the singularity-free workspace of any set of mechanism coordinates. Test cases are included to demonstrate the performance of the method in several situations (Sect. 5.4).

O. Bohigas et al., *Singularities of Robot Mechanisms*,
Mechanisms and Machine Science 41, DOI 10.1007/978-3-319-32922-2_5

5.1 Related Work

Motion planning is a fundamental task in robotics, and there exist numerous techniques to carry it out [4–6], even for mechanisms of a general architecture [7, 8]. To date, however, most works have omitted the treatment of singular configurations, a critical point when closed kinematic chains are present. While strategies exist to traverse these configurations [9, 10], they mainly resort to the mechanism inertia and near a singularity they do not guarantee the position accuracy, nor the equilibrability of the externally-applied forces. To prevent rigidity losses or excessive motor torques it is therefore necessary to avoid these configurations. In commercial robots, this is achieved by fixing joint limits that exclude the presence of singularity, assuming, as a counterpart, a drastic reduction in the available workspace. The use of a planner like the one we propose, on the contrary, allows us to exploit the full motion range of a robot, circumventing singularities as if they were common obstacles to be encircled.

Leaving aside on-line singularity-avoidance methods [11–13], which are meant to locally correct a global pre-computed path, there exist very few algorithms to plan singularity-free motions between distant configurations. Existing strategies are based on deforming a parameterised path between the query configurations [3, 14], reducing the problem to a boundary value problem [15], or treating the singularity locus explicitly as an obstacle [16]. From such planners, a remarkable one is [3], which fastly computes minimal-time trajectories in the DexTAR parallel robot (Fig. 5.1). All of these strategies work well in general, but [14, 15] mention limitations relative to proving path existence in some cases, and the method in [16] is computationally intensive, as it requires constructing polytope approximations of the entire singularity set. In some way or another, likewise, the methods in [3, 14–16] exploit closed-form parameterisations of the C-space, which makes it difficult to extend them to tackle mechanisms with a more complex architecture. In contrast, the method we provide

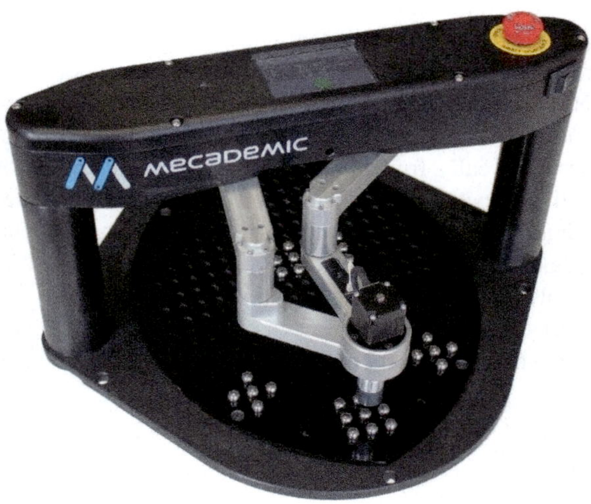

Fig. 5.1 DexTAR, a desktop high-accuracy academic robot [2, 3] used to perform pick-and-place operations. (Courtesy of Prof. Ilian Bonev, ÉTS Montreal.)

makes no recourse to such parameterisations, and can be applied to any nonredundant closed-chain mechanism. As opposed to [16], it treats singularities implicitly, not explicitly as obstacles, resulting in a computationally much less intensive approach.

We should highlight that the proposed method can in particular be used to detect nonsingular transitions between different forward kinematic solutions of a mechanism (also known as assembly modes). The study of these transitions has received ample attention in recent years, since for a long time it was believed that they were impossible [17]. Innocenti and Parenti-Castelli were the first to reveal their existence [18], showing that the movements of a robot can be broader than expected. After this work, several studies have explained and illustrated the transitions in many mechanisms [19–25]. However, their computation has been based on *ad-hoc* methods, or on visual inspection of projections or particular slices of the C-space. The method we present, instead, allows us to find them in general mechanisms in a fairly automatic way, as we showed in [26]. From an input portrait (Sect. 3.4) one can select values of the $\mathbf{v}$ tuple for the different regions, solve Eq. (2.1) to obtain all solutions of the forward kinematic problem, and then use the path planner to check whether a movement exists between one solution to another. Let us see the planner in detail.

5.2 Formulating the Singularity-Free C-Space

Given two configurations of $\mathcal{C} \setminus \mathcal{S}$, $\mathbf{q}_s$ and $\mathbf{q}_g$, our goal is to generate a singularity-free *path* connecting them, i.e. a continuous map

$$\boldsymbol{\kappa} : [0, 1] \rightarrow \mathcal{C} \setminus \mathcal{S},$$

such that $\boldsymbol{\kappa}(0) = \mathbf{q}_s$ and $\boldsymbol{\kappa}(1) = \mathbf{q}_g$. For simplicity, we shall focus on avoding the forward singularities only, because they are those potentially more harmful to the mechanism, but the extension to also deal with inverse singularities is straightforward. Thus, in the rest of the chapter we shall be content in generating paths

$$\boldsymbol{\kappa} : [0, 1] \rightarrow \mathcal{C}_{sfree} = \mathcal{C} \setminus \mathcal{S}_f,$$

where $\mathcal{S}_f$ denotes the forward singularity locus of the mechanism.

Note that $\mathcal{C}_{sfree}$ is the set of points $\mathbf{q} \in \mathcal{C}$ for which

$$det(\mathbf{L}_y) \neq 0, \tag{5.1}$$

where $\mathbf{L}_y$ is the matrix defined in Sect. 2.1. However, continuation methods are aimed at tracing solution sets of systems of equations, not inequalities [27]. To define a path planning strategy based on continuation, therefore, we need to convert Eq. (5.1) into equality form. This can be achieved by simply introducing an auxiliary variable b, and noting that $det(\mathbf{L}_y) \neq 0$ whenever

$$det(\mathbf{L}_y) \cdot b = 1$$

for some value of b. Thus, a system of equations characterising $\mathcal{C}_{sfree}$ is

$$\left.\begin{array}{r}\boldsymbol{\Phi}(\mathbf{q}) = \mathbf{0} \\ det(\mathbf{L}_y) \cdot b = 1\end{array}\right\}. \tag{5.2}$$

We here observe that if inverse singularities need to be avoided, we just need to add $det(\mathbf{L}_z)$ as an additional factor in the second equation, and the rest of the method can be applied without modification.

For convenience, Eq. (5.2) will be written as

$$\mathbf{F}(\mathbf{x}) = \mathbf{0}, \tag{5.3}$$

where

$$\mathbf{x} = (\mathbf{q}, b)$$

and

$$\mathbf{F}(\mathbf{x}) = \begin{bmatrix} \boldsymbol{\Phi}(\mathbf{q}) \\ det(\mathbf{L}_y) \cdot b - 1 \end{bmatrix}. \tag{5.4}$$

Let $\mathcal{M}$ be the set of points $\mathbf{x}$ that satisfy Eq. (5.3), and define the function

$$b(\mathbf{q}) = \frac{1}{det(\mathbf{L}_y(\mathbf{q}))}. \tag{5.5}$$

Observe that the points $\mathbf{x} \in \mathcal{M}$ are in one-to-one correspondence with the points $\mathbf{q} \in \mathcal{C}_{sfree}$, because $\mathbf{q} \in \mathcal{C}_{sfree}$ if, and only if, $\mathbf{x} = (\mathbf{q}, b(\mathbf{q}))$ satisfies Eq. (5.2). Accordingly, all paths in $\mathcal{C}_{sfree}$ are uniquely represented in $\mathcal{M}$, and *vice versa*. Thus, the original problem of computing a path in $\mathcal{C}_{sfree}$ from $\mathbf{q}_s$ to $\mathbf{q}_g$ can be reduced to that of computing a path from $\mathbf{x}_s = (\mathbf{q}_s, b(\mathbf{q}_s))$ to $\mathbf{x}_g = (\mathbf{q}_g, b(\mathbf{q}_g))$ in $\mathcal{M}$. This reduction is advantageous because, by letting the path planner operate in $\mathcal{M}$, rather than in $\mathcal{C}$ directly, guarantees that any obtained path will lie in $\mathcal{C}_{sfree}$ entirely. This eliminates the need of checking singularity crossings in the planner, which, as noted in [14], may be quite difficult due to the intricate structure of the singularity locus.

The correspondence of the two problems is schematically illustrated in Fig. 5.2. The horizontal plane in the bottom represents $\mathcal{C}$, which in this example coincides with the ambient space $\mathcal{Q}$ for simplicity, and the singularity locus $\mathcal{S}_f$ is represented by two red parabolas in this plane. To construct $\mathcal{M}$, we add a new dimension b to $\mathcal{Q}$ (the vertical axis in the figure), and we lift every point $\mathbf{q} \in \mathcal{C}$ to the point $\mathbf{x} = (\mathbf{q}, b(\mathbf{q}))$. Then, $\mathcal{M}$ can be thought of as a new manifold extending infinitely in the direction b, as the projection $\mathbf{q}$ of a point $\mathbf{x} \in \mathcal{M}$ approaches $\mathcal{S}_f$.

Two important observations should be made regarding the search of a path. On the one hand, note that the Jacobian $\mathbf{F}_\mathbf{x}$ has the block structure

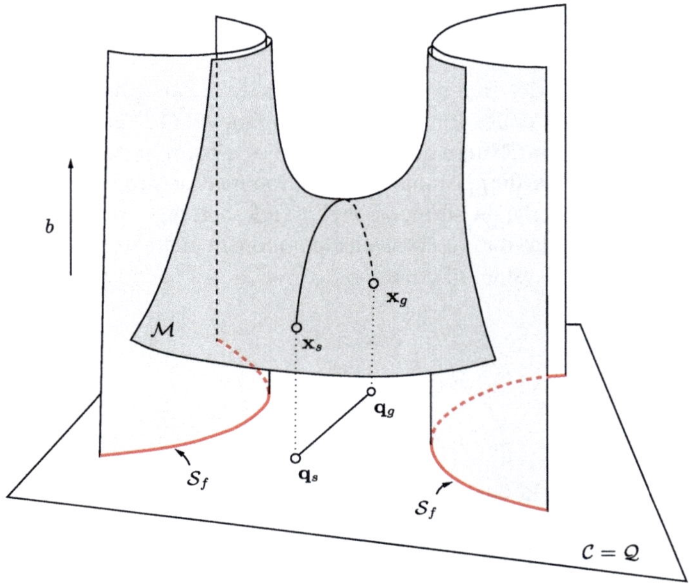

Fig. 5.2 The original problem of computing a singularity-free path in $\mathcal{C}$ connecting $\mathbf{q}_s$ and $\mathbf{q}_g$ is transformed into one of finding an arbitrary path in $\mathcal{M}$ connecting $\mathbf{x}_s$ and $\mathbf{x}_g$. The manifold $\mathcal{M}$ is smooth, which notably simplifies the continuation strategy

$$\mathbf{F_x} = \left[\begin{array}{c|c|c} \mathbf{\Phi_v} & \mathbf{\Phi_y} & \mathbf{0} \\ \hline * & * & det(\mathbf{L}_y) \end{array}\right],$$

from which we see that $\mathbf{F_x}$ is full rank at all points $\mathbf{x} \in \mathcal{M}$. Since $\mathbf{L}_y$ is full rank over all $\mathcal{M}$, $\mathbf{\Phi}_y$ will be full rank too, because both matrices are singular at the same configurations $\mathbf{q}$. Hence, $\mathbf{F_x}$ is never rank deficient in $\mathcal{M}$. By the implicit function theorem this implies that $\mathcal{M}$ has the structure of a smooth manifold [28], which is beneficial from the point of view of numerical continuation [29], since no bifurcations, ridges, or dimension changes are to be found in $\mathcal{M}$. No recourse to branch-switching operations will be necessary therefore [30], which notably simplifies the algorithms to be implemented.

On the other hand, remember that all of the $\mathbf{q}$ coordinates have known bounds in practice, and that those bounds derived from mechanical limits of the joints can always be taken into account by adding new equations to the system (Chaps. 3 and 4). Moreover, $|b|$ should be maintained below a given threshold b_{max} to guarantee some clearance with respect to $\mathcal{S}_f$, determined by the task to accomplish and by the characteristics of the underlying mechanical and control systems. As a result, the search for a path in $\mathcal{M}$ must be restricted to a given domain $\mathcal{D}$ of the $\mathbf{x}$ space, usually defined as the Cartesian product of a number of intervals that correspond to coordinate bounds.

5.3 Exploring the Singularity-Free C-Space

To determine a singularity-free path from $\mathbf{x}_s$ to $\mathbf{x}_g$, we can gradually construct an *atlas* of $\mathcal{M} \cap \mathcal{D}$, i.e., a collection of charts mapping $\mathcal{M} \cap \mathcal{D}$ entirely, where each chart is a local map from $\mathbb{R}^n$ to a neighbourhood of a point in $\mathcal{M}$. The atlas will be computed using a higher-dimensional continuation method proposed by Henderson [29], which generalises the one-dimensional pseudo-arclength procedure [31]. This method generates the atlas for one connected component of the n-dimensional variety implicitly defined by a system of equations

$$\mathbf{F}(\mathbf{x}) = \mathbf{0},$$

where

$$\mathbf{F} : \mathbb{R}^m \rightarrow \mathbb{R}^{m-n}$$

is a differentiable map. In our case, $\mathbf{F}$ is defined as in Eq. (5.4), $m = n_q + 1$, and n is the dimension of $\mathcal{M}$ which, as we have just seen, is a smooth manifold.

We next describe how to generate a full atlas from $\mathbf{x}_s$ (Sects. 5.3.1 and 5.3.2), how to bias the atlas construction towards $\mathbf{x}_g$ for efficiency (Sect. 5.3.3), and, finally, the planner pseudocode (Sect. 5.3.4).

5.3.1 Constructing a Chart

Given a point $\mathbf{x}_i \in \mathcal{M}$, a chart C_i is a diffeomorphism

$$\psi_i : \mathbb{R}^n \rightarrow \mathbb{R}^m$$

from a domain $\mathcal{P}_i \subseteq \mathbb{R}^n$ to an open set of $\mathcal{M}$ containing $\mathbf{x}_i$, such that

$$\psi_i(\mathbf{0}) = \mathbf{x}_i.$$

The map ψ_i is defined using an $m \times n$ matrix $\mathbf{\Psi}_i$ whose columns define an orthonormal basis of $\mathcal{T}_{\mathbf{x}_i}\mathcal{M}$, the n-dimensional tangent space of $\mathcal{M}$ at $\mathbf{x}_i$. This matrix is the one satisfying

$$\begin{bmatrix} \mathbf{F_x}(\mathbf{x}_i) \\ \Psi_i^{\mathsf{T}} \end{bmatrix} \mathbf{\Psi}_i = \begin{bmatrix} \mathbf{0} \\ \mathbf{I}_{n \times n} \end{bmatrix}$$

For a given vector of parameters $\mathbf{s}^i_j \in \mathcal{T}_{\mathbf{x}_i}\mathcal{M}$, the value $\mathbf{x}_j = \psi_i(\mathbf{s}^i_j)$ is determined by computing the point

$$\mathbf{x}^i_j = \mathbf{x}_i + \mathbf{\Psi}_i\, \mathbf{s}^i_j \tag{5.6}$$

of $\mathbb{R}^m$, and then projecting $\mathbf{x}^i_j$ along a ray orthogonal to $\mathcal{T}_{\mathbf{x}_i}\mathcal{M}$, to obtain a point $\mathbf{x}_j \in \mathcal{M}$ (Fig. 5.3a). This projection can be easily achieved by solving

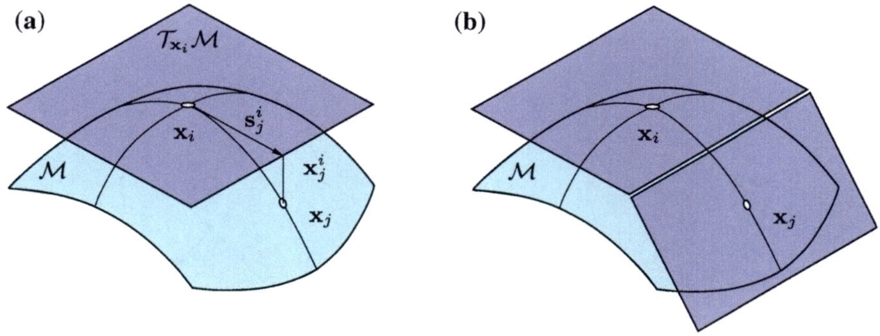

Fig. 5.3 The higher-dimensional continuation method applied to a two-dimensional manifold in $\mathbb{R}^3$. (*Left*) A point $\mathbf{x}_j \in \mathcal{M}$ can be obtained by projecting $\mathbf{x}_j^i$ down to $\mathcal{M}$. (*Right*) If a new chart is defined at $\mathbf{x}_j$, it must be properly coordinated with the chart at $\mathbf{x}_i$, keeping trace of the adjacency relationships

$$\left.\begin{aligned} \mathbf{F}(\mathbf{x}_j) = \mathbf{0} \\ \boldsymbol{\Psi}_i^{\mathsf{T}}\,(\mathbf{x}_j - \mathbf{x}_j^i) = \mathbf{0} \end{aligned}\right\} \tag{5.7}$$

with a Newton-Raphson method initialised at $\mathbf{x}_j^i$ [32], and iteratively updated with the increment $\Delta\mathbf{x}_j$ that fulfills

$$\begin{bmatrix} \mathbf{F}_{\mathbf{x}_i} \\ \boldsymbol{\Psi}_i^{\mathsf{T}} \end{bmatrix} \Delta\mathbf{x}_j = - \begin{bmatrix} \mathbf{F}(\mathbf{x}_i) \\ \boldsymbol{\Psi}_i^{\mathsf{T}}\,(\mathbf{x}_j - \mathbf{x}_j^i) \end{bmatrix}. \tag{5.8}$$

The update is applied until either the norm of the right-hand side of Eq. (5.8) becomes negligible, or a maximum number of iterations is attained. If the process does not converge, the input parameter $\mathbf{s}_j^i$ is assumed to be out of $\mathcal{P}_i$, and the regions of $\mathcal{M}$ not reachable from $\mathcal{P}_i$ will have to be parameterised by other charts (Sect. 5.3.2).

The inverse map of ψ_i, finally, is defined as

$$\mathbf{s}_j^i = \psi_i^{-1}(\mathbf{x}_j) = \boldsymbol{\Psi}_i^{\mathsf{T}}(\mathbf{x}_j - \mathbf{x}_i),$$

and it can be applied to any point $\mathbf{x}_j$ in $\mathbb{R}^m$, regardless of whether it lies in $\mathcal{M}$ or not.

5.3.2 Constructing an Atlas

Due to the curvature of $\mathcal{M}$, the convergence from points of $\mathcal{T}_{\mathbf{x}_i}\mathcal{M}$ to points of $\mathcal{M}$ holds for a small domain $\mathcal{P}_i \subseteq \mathcal{T}_{\mathbf{x}_i}\mathcal{M}$ only. However, since each point in $\mathcal{M}$ is the potential centre of a new chart (Fig. 5.3b), we can always define additional charts

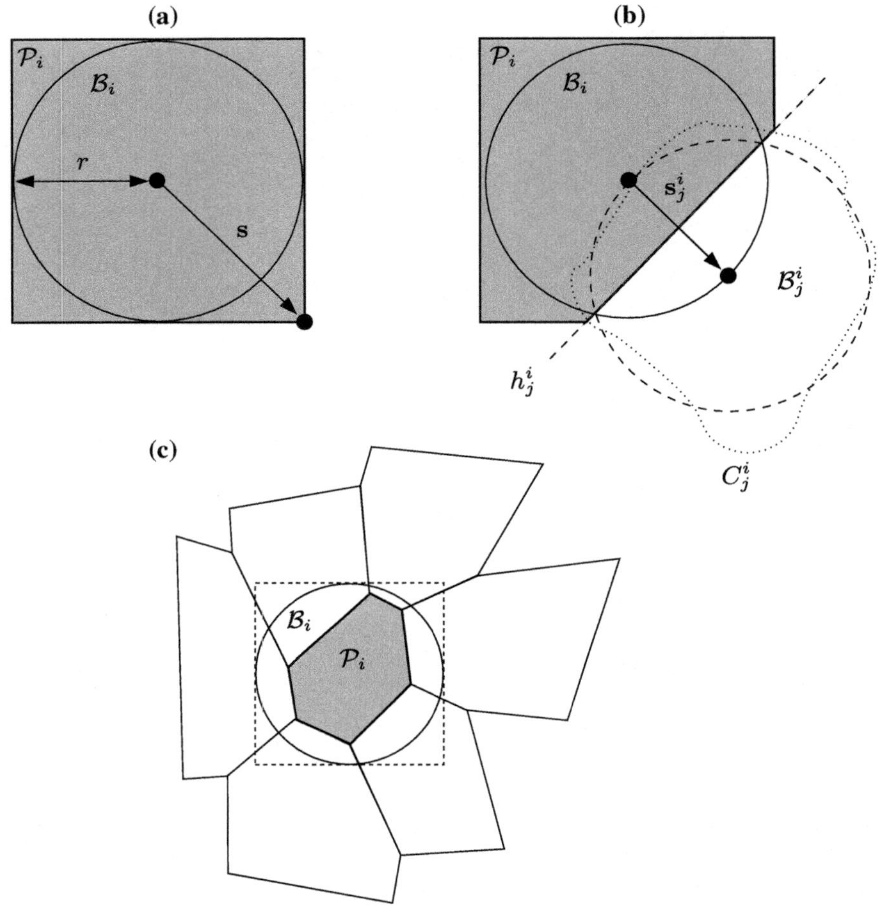

Fig. 5.4 Chart construction. (**a**) Initially, the domain of chart C_i, $\mathcal{P}_i$, is a box circumscribed to a ball of radius r, centred in the origin of $\mathcal{T}_{\mathbf{x}_i}\mathcal{M}$. (**b**) $\mathcal{P}_i$ is trimmed using a ball $\mathcal{B}_j^i$ that approximates C_j^i, the projection on $\mathcal{T}_{\mathbf{x}_i}\mathcal{M}$ of the manifold region covered by C_j. (**c**) When all vertices of $\mathcal{P}_i$ are inside $\mathcal{B}_i$, the chart C_i is said to be *closed*

and coordinate them together to form a whole atlas of $\mathcal{M}$. To this end, we use the method in [29], which proceeds as follows.

For every chart C_i centred in a point $\mathbf{x}_i$ (initially set to $\mathbf{x}_s$), the method initialises the chart domain $\mathcal{P}_i$ to an n-dimensional hypercube circumscribed to a ball $\mathcal{B}_i$ of radius r, both defined on $\mathcal{T}_{\mathbf{x}_i}\mathcal{M}$ (Fig. 5.4a). To add a new chart, the method chooses a vertex of $\mathcal{P}_i$ exterior to $\mathcal{B}_i$, with position vector $\mathbf{s}$, and then finds the point of $\mathcal{B}_i$ in the direction of $\mathbf{s}$, which is given by

$$\mathbf{s}_j^i = \alpha \frac{\mathbf{s}}{\|\mathbf{s}\|} \tag{5.9}$$

with $\alpha = r$. Using $\mathbf{s}_j^i$, the method then determines the point $\mathbf{x}_j^i$ using Eq. (5.6), and attempts to converge to $\mathbf{x}_j = \psi_i(\mathbf{s}_j^i)$ by solving Eq. (5.7). If the Newton-Raphson method converges, and $\mathbf{x}_j$ and $\mathcal{T}_{\mathbf{x}_j}\mathcal{M}$ are close-enough to $\mathbf{x}_j^i$ and $\mathcal{T}_{\mathbf{x}_i}\mathcal{M}$, i.e., if

$$\|\mathbf{x}_j - \mathbf{x}_j^i\| < \epsilon \tag{5.10}$$

and

$$\det(\mathbf{\Psi}_i^{\mathsf{T}}\mathbf{\Psi}_j) > 1 - \epsilon \tag{5.11}$$

for a small threshold ϵ, the method generates a new chart C_j in $\mathbf{x}_j$, with its corresponding domain $\mathcal{P}_j$ initialised to a hypercube. If the method does not converge, C_j is discarded, and a new attempt of chart generation is performed using a smaller value of α in Eq. (5.9). With such a procedure, the area covered by each chart gets naturally adapted to the local curvature of $\mathcal{M}$ in the directions of expansion.

The domain of each new chart C_j has to be properly coordinated with the domains of existing charts in the atlas. To coordinate $\mathcal{P}_i$ and $\mathcal{P}_j$, the method computes a ball $\mathcal{B}_j^i$ that approximates the projection on $\mathcal{T}_{\mathbf{x}_i}\mathcal{M}$ of the manifold region covered by C_j. This projection is indicated as C_j^i in Fig. 5.4b. The method then computes the hyperplane h_j^i defined by the intersection of $\mathcal{B}_i$ and $\mathcal{B}_j^i$, which defines a new face of $\mathcal{P}_i$. The hyperplane eliminates some vertices of $\mathcal{P}_i$ (in particular, the one given by $\mathbf{s}$) and generates new ones. Symmetrically, the polytope $\mathcal{P}_j$ associated with C_j is cropped using an approximation of C_i^j, the projection on $\mathcal{T}_{\mathbf{x}_j}\mathcal{M}$ of the region of $\mathcal{M}$ covered by C_i. The resulting domains $\mathcal{P}_i$ and $\mathcal{P}_j$ are not necessarily continuous in the ambient space $\mathbb{R}^m$ but, under mild conditions, their projections to the manifold slightly overlap, hence covering it continuously. The transition from one local parameterisation to the other can then be achieved by using the direct and inverse maps for the two neighbouring charts. For a given vector of parameters $\mathbf{s}_i$ at the border of $\mathcal{P}_i$, the corresponding vector in $\mathcal{P}_j$ is

$$\mathbf{s}_j = \psi_j^{-1}(\psi_i(\mathbf{s}_i)).$$

When all vertices of $\mathcal{P}_i$ are inside $\mathcal{B}_i$, the chart C_i is said to be *closed*, meaning that no further expansion needs to be attempted from that chart (Fig. 5.4c). Charts whose centre is out of the global domain $\mathcal{D}$ are also considered to be *closed*. The process of chart expansion continues while there are *open* (i.e., nonclosed) charts in the atlas, which occurs when the whole component of $\mathcal{M} \cap \mathcal{D}$ connected to $\mathbf{x}_s$ has been reached.

As an example, Fig. 5.5 illustrates the progression of the algorithm when tracing the surface of a Chmutov surface from a given point, where open and closed charts are coloured in red and blue, respectively. Since this manifold is a surface, the balls $\mathcal{B}_i$ are circles in this case, and the polytopes $\mathcal{P}_i$ are polygons.

It must be said that, to obtain a good coverage of $\mathcal{M}$, r should be chosen such that all charts keep the distance between their tangent space and the manifold below ϵ. In [29], Henderson provides a way to approximate the value of r in terms of the

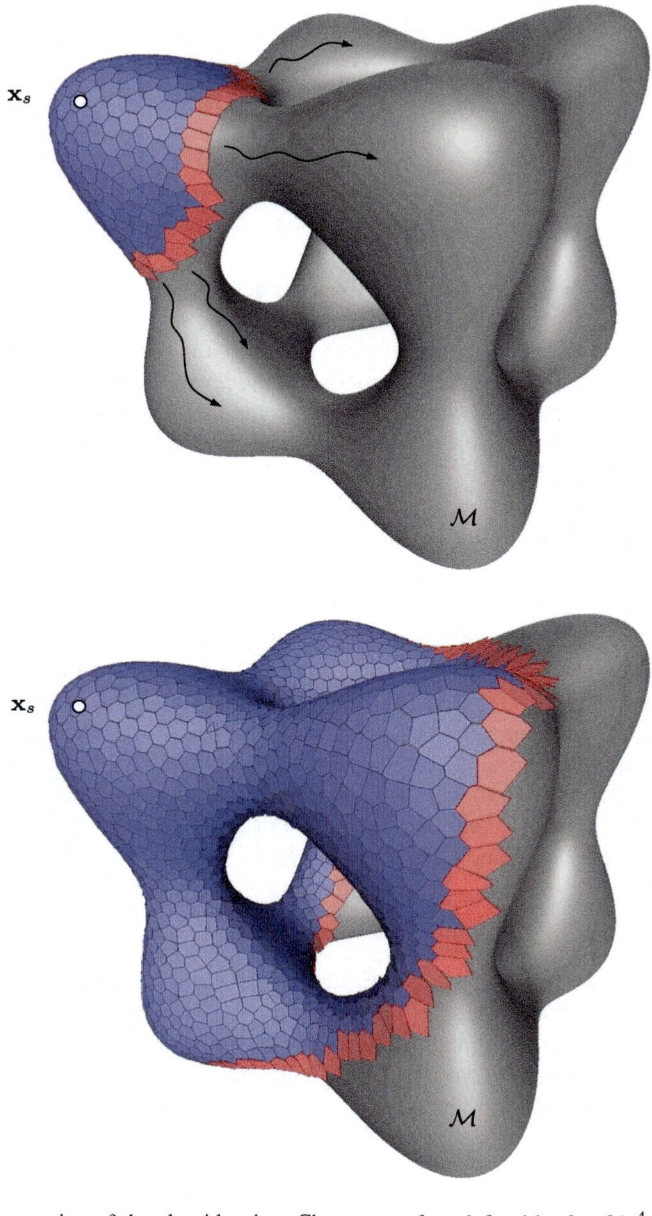

Fig. 5.5 Progression of the algorithm in a Chmutov surface defined by $3 + 8(x^4 + y^4 + z^4) = 8(x^2 + y^2 + z^2)$, using the continuation parameters $r = 0.07$ and $\epsilon = 0.1$. *Red* and *blue* polygons correspond to open and closed charts, respectively. The continuation method progresses in breadth-first order in this example, but the process of chart creation can be biased towards a goal point $\mathbf{x}_g$ if desired, using the A* and Greedy Best-First strategies (see the text for details)

local curvature of the manifold, which ensures that only pieces of size ϵ may remain uncovered when the algorithm finishes. This outcome does not seriously affect the algorithm, but the approximation of r is computationally expensive and it is thus difficult to apply it in practice [33]. As a workaround, Henderson proposes the use of an educated guess when the chart is created, as done in most one-dimensional continuation methods [34], and then let the algorithm adapt the size of the chart through the α parameter, as explained above. Finally note that the norm used to define the balls $\mathcal{B}_i$ in $\mathcal{T}_{\mathbf{x}_i}\mathcal{M}$ does not need to be Euclidean. For instance, the balls may be ellipsoidal if an appropriate norm is used.

5.3.3 Focusing on the Path to the Goal

Once a whole atlas is available, a graph G can be built whose nodes represent the computed charts, and whose edges represent the neighbouring relationships between the charts. If a path from $\mathbf{x}_s$ to $\mathbf{x}_g$ exists, $\mathbf{x}_g$ must be covered by some chart C_k. That is, it must be

$$\psi_k^{-1}(\mathbf{x}_g) \in \mathcal{D}_k \tag{5.12}$$

for some k. A singularity-free path from $\mathbf{x}_s$ to $\mathbf{x}_g$ can then be determined easily by finding a path in G from C_s to C_k, using standard graph search methods. If $\mathbf{x}_g$ is not reachable from any chart in the atlas, path nonexistence is established at the considered value of r.

Note that the previous process computes an exhaustive atlas of the connected component of $\mathbf{x}_s$. This strategy may be worthwhile if we can re-use the atlas a posteriori, e.g. to resolve subsequent planning queries between several configurations, or to determine the full workspace region attainable from $\mathbf{x}_s$. In situations in which we only need to solve a single planning query, however, we can guide the search towards $\mathbf{x}_g$ heuristically, constructing as few charts as possible to reduce the computational burden [7].

A possible approach is to use an A* strategy [35], which only generates the necessary charts to compute a minimum-cost path from $\mathbf{x}_s$ to $\mathbf{x}_g$. At each iteration, this method expands the chart of the point $\mathbf{x}_i$ yielding the lowest estimated cost of the whole movement from $\mathbf{x}_s$ to $\mathbf{x}_g$, while keeping a sorted priority queue of alternative path segments. This cost is the sum of a term

$$G_{ST}(\mathbf{x}_i),$$

which provides the lowest known cost of moving from $\mathbf{x}_s$ to $\mathbf{x}_i$, and a term

$$H_{ST}(\mathbf{x}_i), \tag{5.13}$$

which yields a lower bound of the smallest cost of moving from $\mathbf{x}_i$ to $\mathbf{x}_g$. The value of $G_{ST}(\mathbf{x}_i)$ is maintained during the expansion of the atlas by means of a function

$$C_{ST}(\mathbf{x}_j, \mathbf{x}_k), \tag{5.14}$$

which gives the minimum transition cost between adjacent chart centres. For example, we can use

$$C_{ST}(\mathbf{x}_j, \mathbf{x}_k) = \|\mathbf{x}_j - \mathbf{x}_k\|, \tag{5.15}$$
$$H_{ST}(\mathbf{x}_i) = C_{ST}(\mathbf{x}_i, \mathbf{x}_g), \tag{5.16}$$

which tends to obtain the geodesic path from $\mathbf{x}_s$ to $\mathbf{x}_g$.

The A* strategy typically performs well in manifolds of dimension up to three, but computation times considerably increase in higher dimensions. In such cases, one can simply use a Greedy Best-First strategy, in which the chart C_i to be expanded is just the one yielding a minimum estimated cost $H_{ST}(\mathbf{x}_i)$. In contrast with the A* approach, the Greedy Best-First strategy usually explores a smaller portion of $\mathcal{M}$, but the path obtained may not be close to the optimal one.

Regardless of the strategy adopted, note that any cost function $C_{ST}(\mathbf{x}_j, \mathbf{x}_k)$ can be employed in principle. Depending on the specific context of application, this function may reflect energy consumption, travel distance, or a penalty due to robot collisions. In the latter case, the function only has to assign an infinite cost to the chart-to-chart transitions that are causing a collision [7].

Finally we observe that, since the returned paths use direct motions between adjacent chart centres, they may be slightly jerky, but they can always be smoothed using standard path smoothing techniques [36].

5.3.4 The Planner Algorithm

Algorithm 5.1 gives the pseudocode of the A* strategy described. We omit the pseudocode of the Greedy Best-first strategy because it can be run by calling Algorithm 5.1 with $C_{ST}(\mathbf{x}_j, \mathbf{x}_k) = 0$. Note that, to avoid testing Eq. (5.12) repeatedly, the algorithm creates a chart in $\mathbf{x}_g$ from the outset, and expands the atlas until such a chart is encountered.

At the beginning, the algorithm defines charts at the start and goal configurations and uses them to initialise the atlas (lines 1 to 3). Additionally, it defines the set $\mathcal{H}$ of charts from where the search can be expanded (line 4), and the set $\mathcal{V}$ of closed charts (line 5). It then initialises the pointers to the best parent for each chart (line 6), the cost to reach the initial chart (line 7), and the heuristic function estimating the cost from this chart to the goal (line 8). After that, the algorithm iterates while there are charts in $\mathcal{H}$ and the goal chart has not been reached yet (lines 10–31). In this iteration, the chart C_i with minimum expected cost to the goal is extracted from $\mathcal{H}$ (line 11).

Algorithm 5.1: Singularity-free path planner

PathComputation($\mathbf{F}$, $\mathbf{x}_s$, $\mathbf{x}_g$, H_{ST}, C_{ST}, r, ϵ)

input : The function $\mathbf{F}$ of Eq. (5.3), defining the smooth manifold $\mathcal{M}$; the start and goal configurations, $\mathbf{x}_s$ and $\mathbf{x}_g$, which must be nonsingular; the cost functions H_{ST} and C_{ST} of Eqs. (5.13) and (5.14) used to bias the search towards $\mathbf{x}_g$; the parameters r and ϵ used to construct the atlas.

output: A singularity-free path connecting $\mathbf{x}_s$ and $\mathbf{x}_g$, when one exists, or the empty path $\varnothing$ otherwise.

```
C_s ← NewChart(F, x_s, r)}
C_g ← NewChart(F, x_g, r)}
A ← {C_s, C_g}
H ← {C_s}
V ← ∅
p(s) ← 0
c(s) ← 0
h(s) ← H_ST(x_s, x_g)
C_i ← C_s
while H ≠ ∅ and C_i ≠ C_g do
    C_i ← ExtractMin(H, h)
    if C_i ≠ C_g and c(i) < ∞ then
        while Open(C_i) do
            α ← r
            s ← Vertex(P_i) s.t. s ∉ B_i
            repeat
                C_j ← CreateNeighbourChart(C_i, α, s)
                α ← α · 0.9
            until SimilarCharts(C_i, C_j, ε)
            A ← A ∪ {C_j}
        V ← V ∪ {C_i}
        x_i ← Centre(C_i)
        forall C_j ∈ Neighbour(C_i) do
            if C_j ∉ V then
                x_j ← Centre(C_j)
                t ← c(i) + C_ST(x_i, x_j)
                if C_i ∉ H or t < c(j) then
                    H ← H ∪ {C_j}
                    p(j) ← i
                    c(j) ← t
                    h(j) ← t + H_ST(x_j, x_g)
if C_i = C_g then
    Return(ReconstructPath(s, g, p))
else
    Return(∅)
```

If C_i is not the goal chart (line 12), and while it is not a closed chart (line 13), all of its neighbours are generated (lines 14–20). Charts whose centre is out of the domain $\mathcal{D}$ are considered closed and, thus, their neighbours are not generated. If necessary, the neighbours are generated by selecting a vertex of $\mathcal{P}_i$ not in $\mathcal{B}_i$, and using this vertex to define the parameters as in Eq. (5.9). Chart creation is attempted until the conditions given in Eqs. (5.10) and (5.11) are met (line 19). When this happens, the new chart is added to the atlas, coordinating it with the charts already present in it (line 20). Once all the neighbours for C_i are generated, the chart is added to $\mathcal{V}$ (line 21) and the search is expanded from it. For each of the open neighbours, the tentative cost to the neighbour via C_i is computed (line 26). Charts not yet in $\mathcal{H}$, or charts where the tentative cost is lower than the best cost computed up to the moment, are added to $\mathcal{H}$ (line 28), changing their parent chart (line 29), setting their new cost (line 30), and finally computing the heuristic estimation of the cost to the goal (line 31). At the end of the search, if the goal was found, the path connecting $\mathbf{x}_s$ and $\mathbf{x}_g$ is derived using the pointers to the parent chart stored in p (line 33). Otherwise, an empty path is returned (line 35), indicating that it is not possible to determine a singularity-free path at the considered resolution.

The cost of the algorithm at each step is dominated by the cost of two searches among the set of charts: one to find the potential neighbours of a new chart when adding it to the atlas (line 20), and another one to find an open chart from which to continue the search. The performance of the first search can be increased using a k-d tree storing the centres of the charts. If $\mathcal{H}$ is implemented using a heap, both the extraction of the next chart to expand (line 11) and the insertion of a new chart in $\mathcal{H}$ (line 28) are logarithmic in the number of expandable charts.

5.4 Case Studies

The performance of the A* planner is next illustrated in two situations: first in a fictitious three-dimensional C-space, and then in a physical 3-RRR parallel robot. The former case is chosen for its simplicity, to illustrate and visualise the method in detail, and the latter shows the method's performance on a real application of considerable complexity. Note that in both cases no use is made of closed-form parameterisations of $\mathcal{C}$. Also, despite the singularity locus and the workspace are shown for reference in the figures (computed using the methods in Chaps. 3 and 4), explicit knowledge of such sets is not used by the planner in any form. All results reported have been obtained with an implementation in C of the A* method available in [37], executed on a MacBook Pro equipped with a 2.66 GHz Intel Core i7 processor. We have used the cost functions in Eqs. (5.15) and (5.16) everywhere.

5.4.1 A Simple Example

Consider the fictitious C-space defined implicitly by

$$\mathbf{\Phi}(q_1, q_2, q_3) = q_1 - \sigma \cos(\omega\,(q_2^2 + q_3^2)) = 0,$$

with $\sigma = 0.5$ and $\omega = 0.25$. This equation defines a sinusoidal surface in the $\mathbf{q} = (q_1, q_2, q_3)$ space, which we show in Fig. 5.6 for

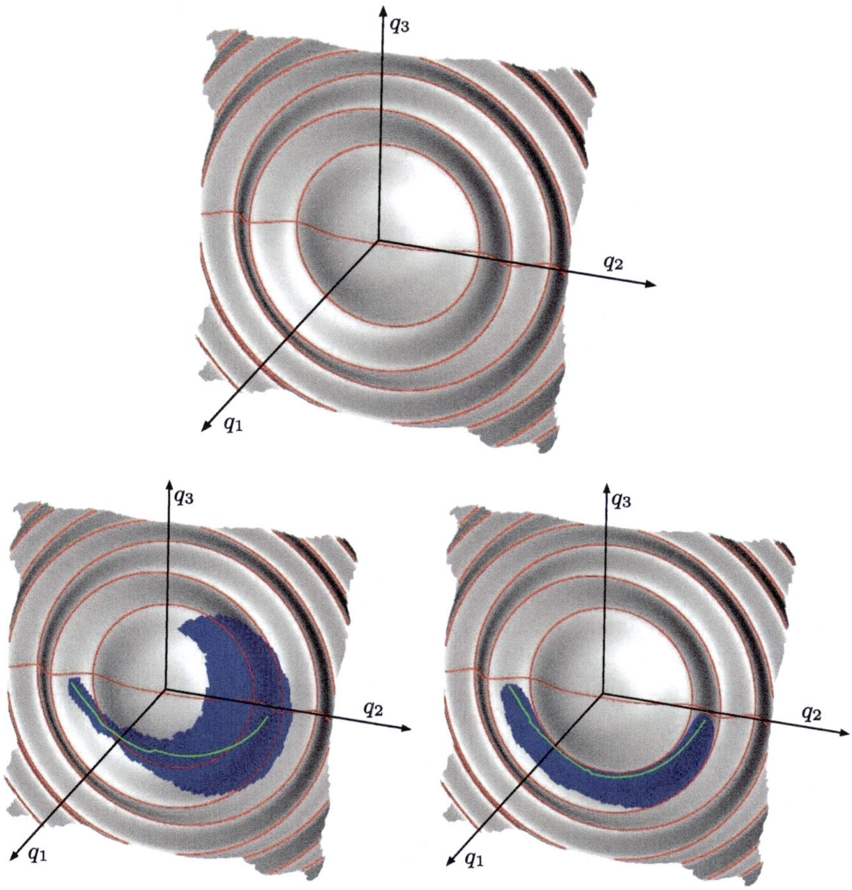

Fig. 5.6 (*Top*) A fictitious three-dimensional C-space with its singularities highlighted in *red*, assuming that q_1 and q_2 are the actuated degrees of freedom. (*Bottom*) The path computed by the proposed algorithm when neglecting and considering singularity avoidance (*left* and *right*, respectively). In the plots, the singularity locus is highlighted in *red*, the charts explored to connect the two query configurations are shown as *blue* polygons, and the final returned path is shown in *green*. While the path on the *left* figure crosses the singularity set twice, the path on the *right* figure is singularity-free

$$\mathbf{q} \in [-1, 1] \times [-20, 20] \times [-20, 20].$$

Let us assume that the tuple of actuated degrees of freedom is $\mathbf{v} = (q_1, q_2)$, so that $\mathbf{y} = (q_3)$. Then, the forward singularities are given by the equation

$$det(\mathbf{\Phi_y}) = \frac{\partial \mathbf{\Phi}}{\partial q_3} = 2\,\sigma\,\omega\,q_3\,\sin(\omega\,(q_2^2 + q_3^2)) = 0,$$

which holds whenever $\omega\,(q_2^2 + q_3^2) = n\,\pi$, with $n \in \mathbb{Z}$, or when $q_3 = 0$. Thus the singularity locus is formed by a family of concentric circles and a sinusoidal curve, which are the red curves shown in Fig. 5.6 (top). Note that, in accordance with the geometric interpretation of Sect. 2.2, the points of the red locus are those where the tangent plane of the C-space projects vertically to a line of the (q_1, q_2) plane.

Figure 5.6 (bottom) shows the results obtained by the A* planner, when trying to connect the configurations

$$\mathbf{q}_s = (0, 4.33, -0.38)$$

and

$$\mathbf{q}_g = (0, -4.33, -0.38).$$

To compare the results, the figure shows the computed path, in green, when navigating $\mathcal{C}$ (left figure) and $\mathcal{M}$ (right figure), so that the crossing of singularities is permitted and avoided, respectively. In both cases the planner returns the shortest path up to the resolution of the generated atlas. The charts of these atlases are shown in blue in both figures.

As we see, the path on the left figure crosses the singularity locus twice, while the path on the right figure, although longer, avoids crossing it. While the latter path approaches the singularity locus, a minimum clearance is always guaranteed because the algorithm keeps the value of $|b|$ below a threshold b_{max}. In this case we used $b_{max} = 12$, resulting in the shown path, but alternative paths with a larger clearance can be obtained if desired. In any case, b_{max} should always be larger than the maximum of $|b(\mathbf{q}_s)|$ and $|b(\mathbf{q}_g)|$, so as to guarantee that the domain $\mathcal{D}$ within which the search is confined (Sect. 5.2) includes both $\mathbf{x}_s$ and $\mathbf{x}_g$. The computation of the singularity-free path took 0.04 seconds in this example, using the continuation parameters $r = 0.25$ and $\epsilon = 0.25$.

5.4.2 A 3-$\underline{R}$RR Parallel Robot

The planner has also been tested experimentally in the 3-$\underline{R}$RR robot of Fig. 5.7. The full system features a graphical interface to select start and goal configurations, analyse sensory data, and to plan, simulate and execute motion paths. The robot is controlled by means of an Arduino MEGA board, which implements the lower-level control loops, and acquires all of the sensory data to return it to the graphical interface [38].

Fig. 5.7 A 3-RRR parallel robot developed to test the planner proposed

The robot is in fact a physical version of the 3-RRR mechanism studied in Sect. 3.5.1 (see the section and Fig. 3.6 for notation). The active joints are those connected to the base triangle, and the geometric parameters are the same as those indicated in Table 3.2, corresponding to a robot studied by Bonev in [39, 40]. Due to its relatively large workspace, and the fact that mounting the actuators on the base reduces weight in the mobile equipment, this is a quite common architecture for a 3-DOF planar parallel mechanism [40–42].

We set the geometry of the manipulator so that the two links of each leg move in different planes and thus they cannot collide. In this way we permit a change of working mode of the leg, which illustrates that such a change can be done without losing control of the robot. Doubts about this fact were initially expressed in [40, p. 72]. Springs in intermediate leg joints or the exploitation of link inertias were said

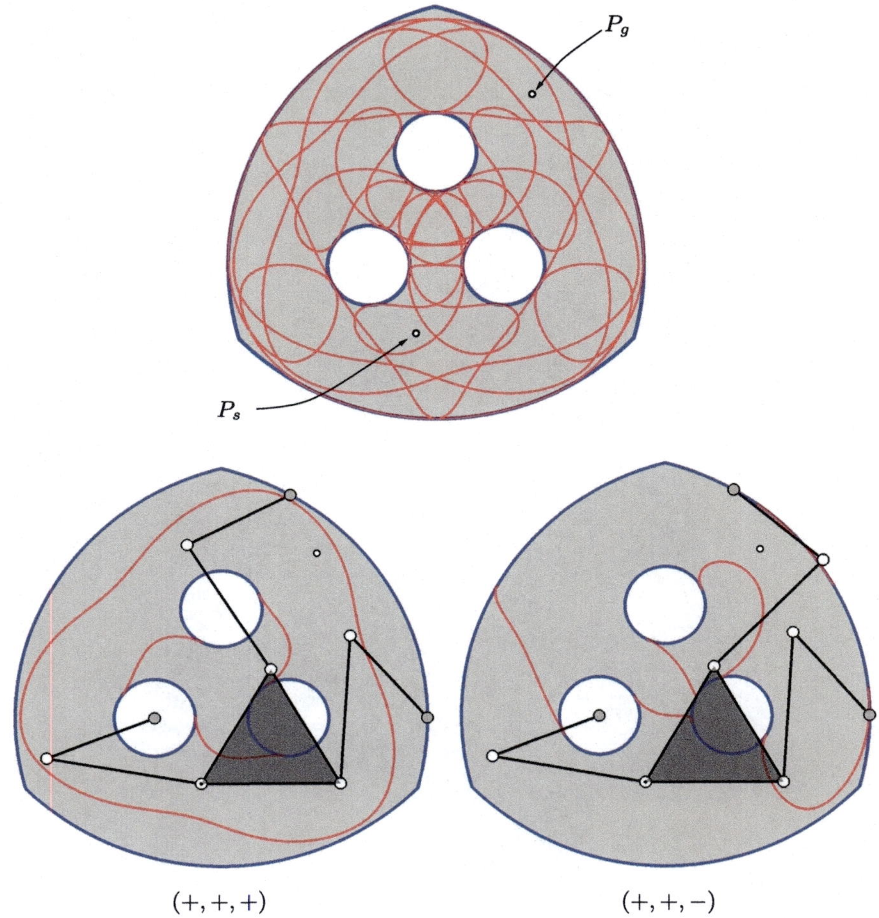

Fig. 5.8 (*Top*) The constant orientation workspace of the 3-RRR robot for $\theta_8 = 0$ (in gray), using $P_7 = (x_7, y_7)$ as a reference point (see Fig. 3.6). The forward and inverse singularities are shown in red and blue, respectively. (*Bottom*) Workspace "layers" relative to two working modes, and their singularity curves

to be necessary to force the flipping of the legs, but the truth is that such flippings occur naturally and in a controlled manner when actuating the A_i joints, as long as the movement avoids the forward singularities. In fact, the possibility of switching working modes through active-joint control was later demonstrated using real prototypes [43].

The same formulation of Sect. 3.5.1 can be used here to write Eq. (5.2). The singularity locus of this mechanism is two-dimensional, and its shape is rather complex. Some of its slices for a fixed value of θ_8 have been shown in Fig. 3.10, and the slice corresponding to $\theta_8 = 0$ is reproduced in Fig. 5.8 (top) for convenience. From the

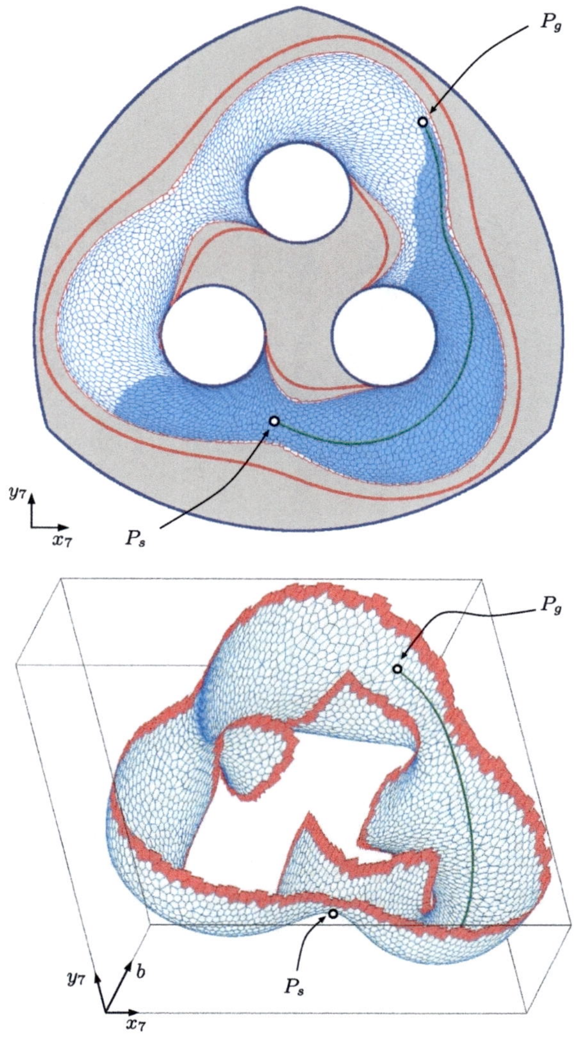

Fig. 5.9 A computed singularity-free path that connects P_s to P_g, assuming $\theta_8 = 0$. All results are shown projected to the (x_7, y_7) plane (*top*) and to the (x_7, y_7, b) space (*bottom*). The path is shown in *green* overlaid on the atlas of the singularity-free component of $\mathcal{C}$ attainable from P_s. The atlas charts are shown coloured in *white* or *red* depending on whether they lie inside or outside of the domain $\mathcal{D}$. The region actually explored to connect P_s and P_g is shown in *blue* in the *top* figure

figure, it may seem that the safe motion range of the mechanism is severely reduced by the presence of singularities. For instance, it appears that P_7 in Fig. 3.6 cannot be moved from P_s to P_g while keeping $\theta_8 = 0$. Note however that we are actually observing a projection of the C-space to the (x_7, y_7) plane, and that for each pose (x_7, y_7, θ_8) of the platform there are up to eight inverse kinematic solutions of the mechanism. Each such solution corresponds to a different working mode, identified by the sign triple $(\kappa_1, \kappa_2, \kappa_3)$, where κ_i gives the orientation of the leg triangle $P_i P_{i+3} P_{i+6}$ [40]. The C-space, thus, has more structure than it looks. It is formed by

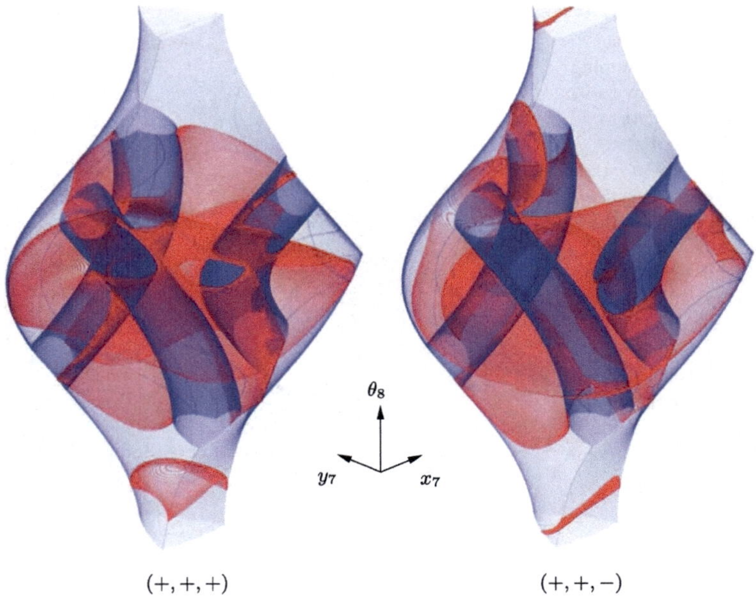

Fig. 5.10 Workspace boundaries of the 3-RRR robot of Fig. 5.7 (in *blue*) and its forward singularities (in *red*) shown in the full (x_7, y_7, θ_8) space for the $(+, +, +)$ and $(+, +, -)$ working modes. The orientation of the axes is shown for reference in a random position. The plots span the whole range $[-\pi, \pi]$ in θ_8. The companion web page of this book offers animated versions of these figures [44]

several "layers", each one of them corresponding to a different working mode, and if we project such layers separately larger singularity-free regions become revealed.

For example, the layers corresponding to the $(+, +, +)$ and $(+, +, -)$ working modes for $\theta_8 = 0$ are shown in Fig. 5.8 (bottom), together with the portion of the singularity curve lying on them, and a representative configuration in each case, with P_7 coinciding on P_s. As we see, P_s and P_g are actually connectable through a singularity-free path lying entirely in the $(+, +, +)$ layer. The A* planner is able to compute such a path in only 0.05 seconds, using $b_{max} = 3.333$, $r = 0.2$, and $\epsilon = 0.25$, obtaining the results shown in Fig. 5.9, where the positions $\mathbf{p}_s = (0.4, -0.6)$ and $\mathbf{p}_g = (1.4, 1.5)$ have been assumed for P_s and P_g. Note from the bottom plot how $\mathcal{M}$ reaches higher values of b as it approaches the singularities, but the found path avoids these zones and the algorithm returns the shortest possible path in $\mathcal{M}$ (a video in [44] shows different views of this 3-D plot). The partial atlas generated to resolve the planning query is shadowed in blue in the top plot.

Constant-orientation paths changing from one layer to another can be obtained as well (the change occurring when P_7 reaches a blue curve), but the full potential of the planner comes out when, to reach a goal, the platform is forced to rotate and change its working mode along the path. To have an idea of the difficulty, Fig. 5.10 shows the full workspace boundaries in the (x_7, y_7, θ_8) coordinates (in blue), and the

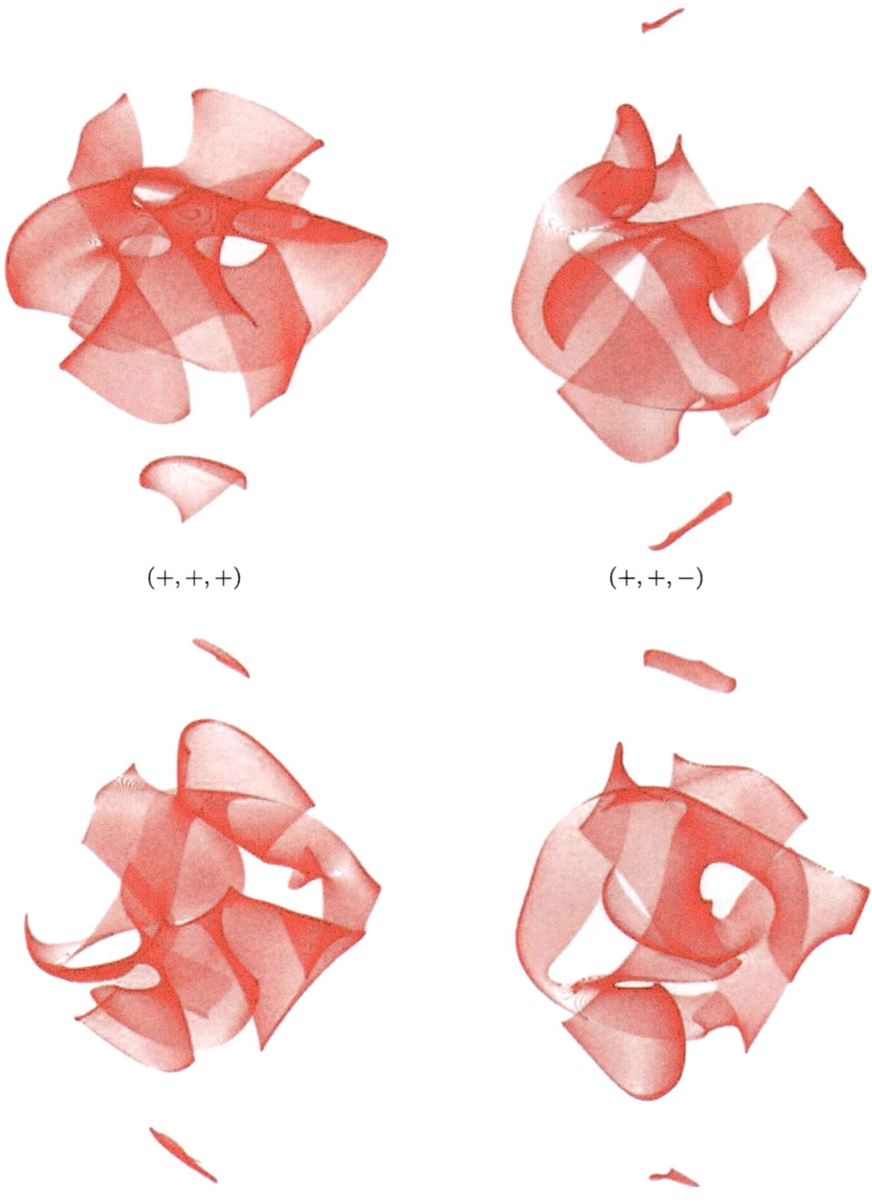

Fig. 5.11 The forward singularity surfaces of the 3-RRR robot in the (x_7, y_7, θ_8) space, for the working modes $(+,+,+)$, $(+,+,-)$, $(+,-,+)$, and $(+,-,-)$

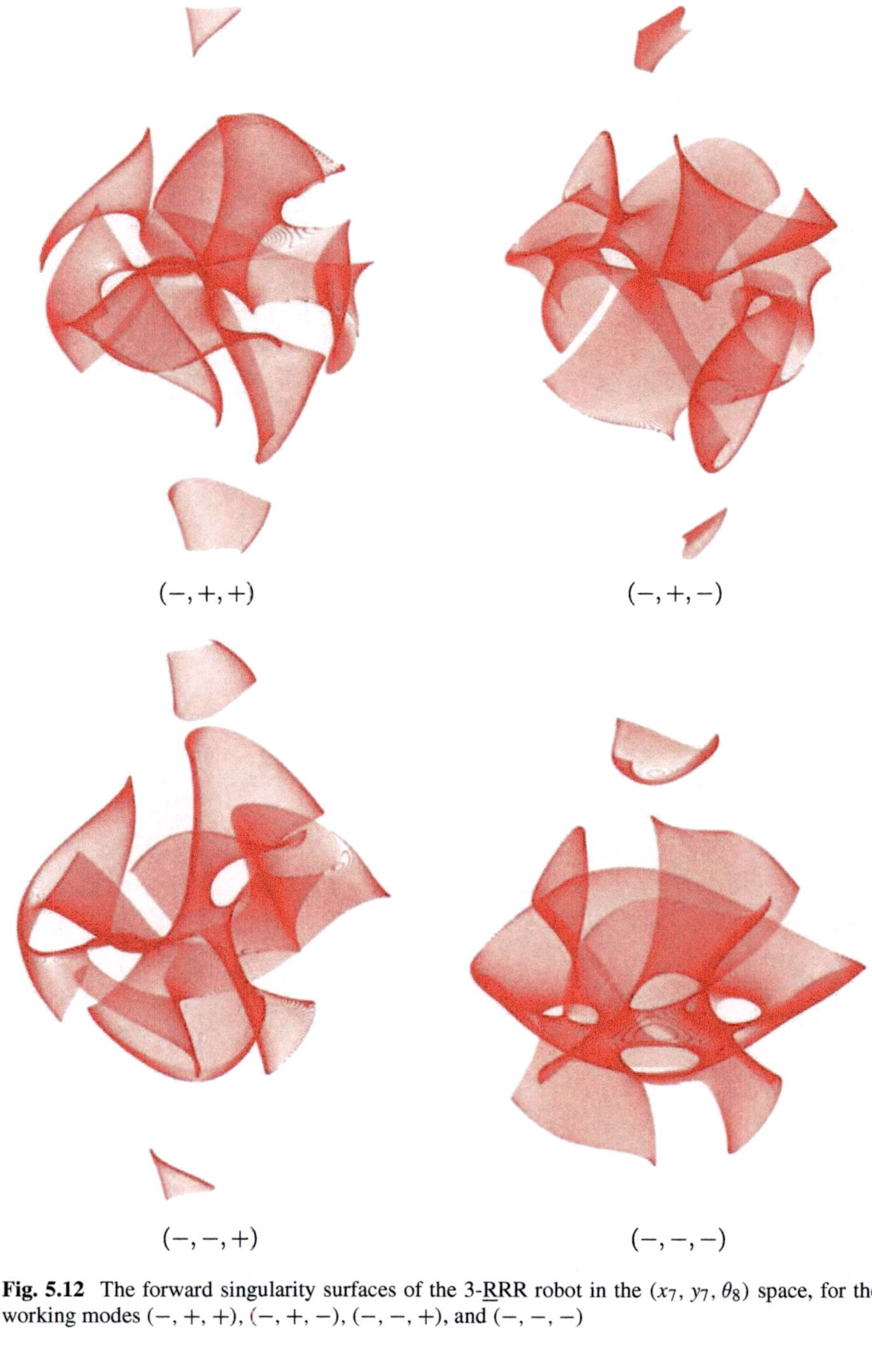

Fig. 5.12 The forward singularity surfaces of the 3-RRR robot in the (x_7, y_7, θ_8) space, for the working modes (−, +, +), (−, +, −), (−, −, +), and (−, −, −)

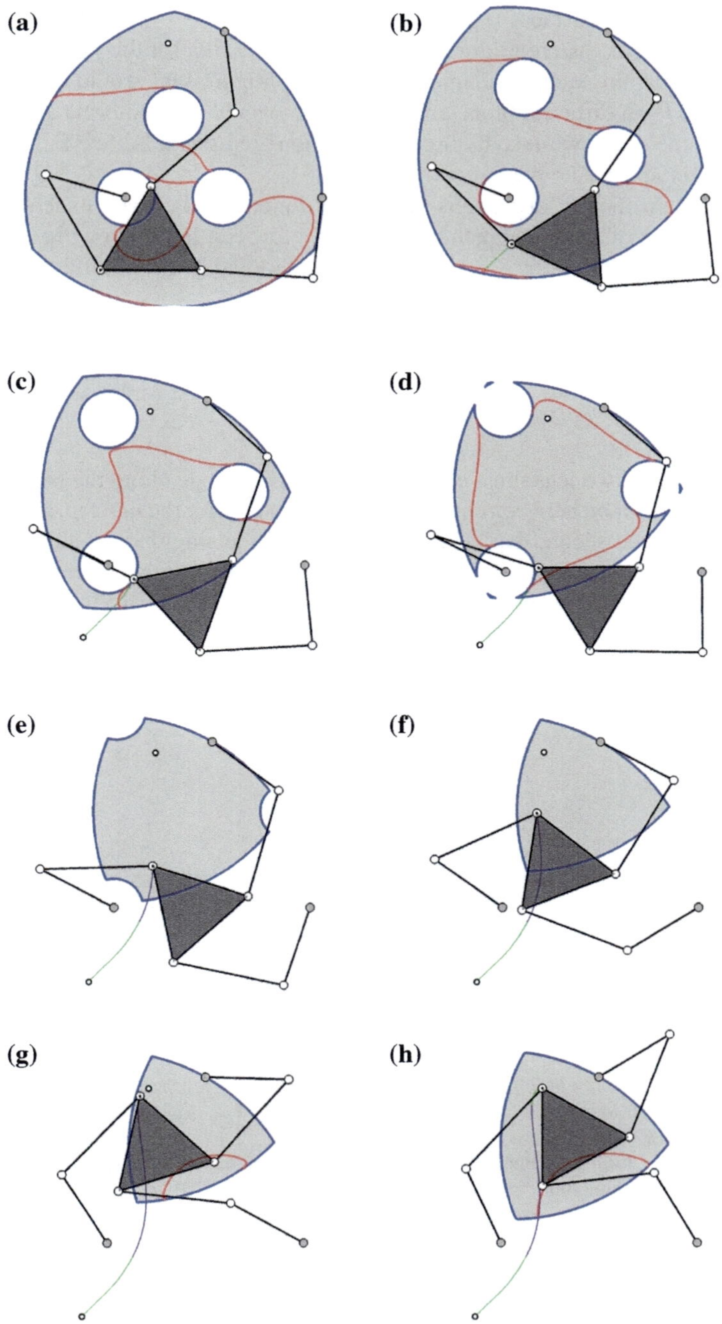

Fig. 5.13 Several instants of the movement of the robot in Fig. 5.7 when executing the computed path between configurations (**a**) and (**h**). See [44] for a full video

forward singularity surfaces (in red), for the modes $(+, +, +)$ and $(+, +, -)$. The preface of the book also reproduces the corresponding plot for the $(-, -, -)$ mode. For a given planning query, the initial and final configurations would correspond to positions of P_7 in different plots, and the planner would have to bring P_7 from one location to the other, while avoiding the capricious red surfaces! See Figs. 5.11 and 5.12 to better appreciate them.

Due to the difficulty of visualising a three-dimensional curve in such plots, we show the results of a planned path by means of a few configurations (Fig. 5.13). The first and last configurations, (a) and (i), are the start and goal ones to be connected, which are given by

Configuration	P_7	θ_8	Working mode
Start	$(-0.3, -0.9)$	0	$(+, -, -)$
Goal	$(0.5, 1.9)$	$-\frac{\pi}{2}$	$(-, +, -)$

Note that at least two legs must change their working mode along the path, and this can be seen to happen between configurations (c) and (d) for the first leg, and between (f) and (g) for the second leg. The workspace and the singularities shown in each picture correspond to the orientation of the moving triangle at that moment only (i.e., they are horizontal cross-sections of plots analogous to those in Fig. 5.10) and can be seen to vary with the orientation. The computation of this path took 1 second in this example, using $r = 0.3$ and $\epsilon = 0.25$. A video corresponding to Fig. 5.13 can be seen in the companion web page of this book [44].

References

1. Q. Jiang, C.M. Gosselin, Determination of the maximal singularity-free orientation workspace for the Gough-Stewart platform. Mech. Mach. Theory **44**(6), 1281–1293 (2009)
2. MECADEMIC: Affordable Desktop High-Accuracy Robot Arms, http://www.mecademic.com. Accessed 26 Dec 2015
3. F. Bourbonnais, P. Bigras, I. Bonev, Minimum-time trajectory planning and control of a pick-and-place five-bar parallel robot. IEEE/ASME Trans. Mechatron. **20**(2), 740–749 (2015)
4. J. Canny, *The Complexity of Robot Motion Planning* (The MIT Press, Cambridge, Massachusetts, 1988)
5. J.-C. Latombe, *Robot Motion Planning* (Kluwer Academic Publisher, 1991)
6. L.E. Kavraki, S.M. LaValle, *Springer Handbook of Robotics*, ch. Motion Planning (Springer, 2008), pp. 109–131
7. J.M. Porta, L. Jaillet, O. Bohigas, Randomized path planning on manifolds based on higher-dimensional continuation. Int. J. Robot. Res. **31**(2), 201–215 (2012)
8. J. Porta, L. Ros, O. Bohigas, M. Manubens, C. Rosales, L. Jaillet, The CUIK suite: motion analysis of closed-chain multibody systems. IEEE Robot. Autom. Mag. **21**(3), 105–114 (2014)
9. C. Jui, Q. Sun, Path tracking of parallel manipulators in the presence of force singularity. ASME J. Dyn. Syst. Measur. Control **127**, 550–563 (2005)
10. S. Briot, V. Arakelian, On the dynamic properties and optimum control of parallel manipulators in the presence of singularity, in *Proceedings of the IEEE International Conference on Robotics and Automation, ICRA (Pasadena, USA)* (2008), pp. 1549–1555

11. S. Bhattacharya, H. Hatwal, A. Ghosh, Comparison of an exact and an approximate method of singularity avoidance in platform type parallel manipulators. Mech. Mach. Theory **33**(7), 965–974 (1998)
12. H. Schaub, J.L. Junkins, Singularity avoidance using null motion and variable-speed control moment gyros. J. Guidance Control Dyn. **23**(1), 11–16
13. G. Marani, J. Kim, J. Yuh, W.K. Chung, A real-time approach for singularity avoidance in resolved motion rate control of robotic manipulators, in *Proceedings of the IEEE International Conference on Robotics and Automation, ICRA (Washington D.C., USA)*, vol. 2 (2002), pp. 1973–1978
14. B. Dasgupta, T.S. Mruthyunjaya, Singularity-free path planning for the Stewart platform manipulator. Mech. Mach. Theory **33**(6), 711–725 (1998)
15. S. Sen, B. Dasgupta, A.K. Mallik, Variational approach for singularity-free path-planning of parallel manipulators. Mech. Mach. Theory **38**(11), 1165–1183 (2003)
16. A.K. Dash, I. Chen, S.H. Yeo, G. Yang, Workspace generation and planning singularity-free path for parallel manipulators. Mech. Mach. Theory **40**(7), 776–805 (2005)
17. K.H. Hunt, E.J.F. Primrose, Assembly configurations of some in-parallel-actuated manipulators. Mech. Mach. Theory **28**(1), 31–42 (1993)
18. C. Innocenti, V. Parenti-Castelli, Singularity-free evolution from one configuration to another in serial and fully-parallel manipulators. ASME J. Mech. Des. **120**, 73 (1998)
19. P.R. McAree, R.W. Daniel, An explanation of never-special assembly changing motions for 3–3 parallel manipulators. Int. J. Robot. Res. **18**(6), 556–574 (1999)
20. E. Macho, O. Altuzarra, C. Pinto, A. Hernandez, Transitions between multiple solutions of the direct kinematic problem, eds. by J. Lenarcic, P. Wenger. *Advances in Robot Kinematics: Analysis and Design* (Springer, 2008), pp. 301–310
21. M. Zein, P. Wenger, D. Chablat, Singular curves in the joint space and cusp points of 3-RPR parallel manipulators. Robotica **25**(6), 717–724 (2007)
22. M. Urízar, V. Petuya, O. Altuzarra, E. Macho, A. Hernández, Computing the configuration space for tracing paths between assembly modes. J. Mech. Robot. **2**(3), 031002–1–031002–11 (2010)
23. S. Caro, P. Wenger, D. Chablat, Non-singular assembly mode changing trajectories of a 6-DOF parallel robot, in *Proceedings of the ASME International Design Engineering Technical Conferences and Computers and Information in Engineering Conference, IDETC/CIE (Chicago, USA)* (2012), pp. 1–10
24. M. Urízar, V. Petuya, O. Altuzarra, A. Hernández, Assembly mode changing in the cuspidal analytic 3-RPR. IEEE Trans. Robot. **28**(2), 506–513 (2012)
25. M. Manubens, G. Moroz, D. Chablat, P. Wenger, F. Rouillier, Cusp points in the parameter space of degenerate 3-RPR planar parallel manipulators. J. Mech. Robot. **4**(4), 041003–1–041003–8 (2012)
26. O. Bohigas, M.E. Henderson, L. Ros, J.M. Porta, A singularity-free path planner for closed-chain manipulators, in *Proceedings of the IEEE International Conference on Robotics and Automation, ICRA (St. Paul, USA)* (2012), pp. 2128–2134
27. M.E. Henderson, *Numerical Continuation Methods for Dynamical Systems: Path Following and Boundary Value Problems*, ch. Higher-Dimensional Continuation (Springer, 2007), pp. 77–115
28. S.G. Krantz, H.R. Parks, *The Implicit Function Theorem: History*, Theory and Applications (Birkhäuser, 2002)
29. M.E. Henderson, Multiple parameter continuation: computing implicitly defined k-manifolds. Int. J. Bifurcat. Chaos **12**(3), 451–476 (2002)
30. M.E. Henderson, Multiparameter parallel search branch switching. Int. J. Bifurcat. Chaos Appl. Sci. Eng. **15**(3), 967–974 (2005)
31. H.B. Keller, *Applications of Bifurcation Theory* (Academic Press, 1977)
32. W.C. Rheinboldt, MANPACK: a set of algorithms of computations on implicitly defined manifolds. Comput. Math. Appl. **32**(12), 15–28 (1996)

33. Q. Zhang, G. Xu, Curvature computations for n-manifolds in $\mathbb{R}^{n+m}$ and solution to an open problem proposed by R. Goldman. Comput. Aided Geom. Des. **24**(2), 117–123 (2007)
34. W.J.F. Govaerts, *Numerical Methods for Bifurcations of Dynamical Equilibria* (Society for Industrial and Applied Mathematics (SIAM), 2000)
35. S.J. Russell, P. Norvig, *Artificial Intelligence: A Modern Approach* (Prentice Hall, 2003)
36. D. Berenson, S.S. Srinivasa, J.J. Kuffner, Task space regions: a framework for pose-constrained manipulation planning. Int. J. Robot. Res. **30**(12), 1435–1460 (2011)
37. The CUIK Project Home Page, http://www.iri.upc.edu/cuik. Accessed 16 Jun 2016
38. A. Rajoy, Planificación y ejecución de trayectorias libres de singularidades en robots paralelos 3-$\underline{R}$RR, Master's thesis, Universitat Politècnica de Catalunya, 2015. Available through http://goo.gl/Hsba1K. Accessed 26 Dec 2015
39. I.A. Bonev, C.M. Gosselin, Singularity loci of planar parallel manipulators with revolute joints," in *Proceedings of the 2nd Workshop on Computational Kinematics (Seoul, South Korea)* (2001), pp. 291–299
40. I.A. Bonev, *Geometric Analysis of Parallel Mechanisms*. PhD thesis, Faculté des Sciences et de Génie, Université de Laval, 2002
41. J. Kotlarski, B. Heimann, T. Ortmaier, Influence of kinematic redundancy on the singularity-free workspace of parallel kinematic machines. Front. Mech. Eng. **7**(2), 120–134 (2012)
42. A. Peidró, O. Reinoso, A. Gil, J.M. Marín, L. Payá, A virtual laboratory to simulate the control of parallel robots. *IFAC-PapersOnLine*, vol. 48, no. 29, pp. 19–24, 2015. Paper from the IFAC Workshop on Internet Based Control Education (IBCE15), Brescia (Italy)
43. L. Campos, F. Bourbonnais, I.A. Bonev, P. Bigras, Development of a five-bar parallel robot with large workspace, in *Proceedings of the ASME International Design Engineering Technical Conferences and Computers and Information in Engineering Conference, IDETC/CIE (Montreal, Canada)* (2010)
44. Companion web page of this book, http://www.iri.upc.edu/srm. Accessed 16 Jun 2016

Chapter 6
Planning with Further Constraints

In the preceding chapter we have measured the clearance of the path relative to the singularity locus in terms of the determinant of a Jacobian matrix. It is often argued that such a measure lacks physical significance [1], but in the early design stages of the mechanism only the kinematic architecture has been decided. Further information leading to more meaningful measures, such as the desired positioning accuracy of the end effector, the maximum force/velocity deliverable by the actuators, or the link inertias, is often not yet available. The planner of Chap. 5 already covers the needs of a robot designer in such situations, because it can compute an accurate roadmap of the potentially-feasible movements. However, for subsequent design stages, or for motion planning during normal operation, further physical limitations should be taken into account.

This chapter presents a way to do so exemplified in the particular case of rigid-limbed and cable-driven hexapods. Following the spirit of [2, 3], the clearance of the path is the one that results from only allowing wrench-feasible configurations, i.e. those in which the leg forces remain within given limits, for any platform wrench subject to set-bounded ellipsoidal uncertainty in all directions (Sect. 6.1).

The method relies on defining a system of equations whose solution manifold corresponds to the wrench-feasible C-space of the mechanism, so that maneuvering through such manifold guarantees singularity avoidance at all times while maintaining leg forces and lengths within their permitted bounds (Sect. 6.2). This manifold, as well as any of its generic submanifolds obtained by constraining some of the pose parameters, are shown to be smooth everywhere (Sect. 6.3), which allows us to exploit the same continuation strategy of Chap. 5 to derive paths between given configurations. The approach is illustrated by means of several cable-driven and rigid-limbed hexapods (Sect. 6.4), and hints on how to compute the uncertainty ellipsoid and on the treatment of alternative constraints and architectures are given in the end (Sects. 6.5 and 6.6).

O. Bohigas et al., *Singularities of Robot Mechanisms*,
Mechanisms and Machine Science 41, DOI 10.1007/978-3-319-32922-2_6

6.1 Wrench-Feasibility Constraints

For the purpose of this chapter, a hexapod consists of a moving platform connected to a fixed base by means of six legs, which can be universal-prismatic-spherical chains, as in Gough-Stewart platforms (Sect. 3.5.2), or cables winding around independent winches, as in cable-driven robots (Fig. 6.1). The leg lengths d_i can be varied within prescribed limits $(\underline{d_i}, \overline{d_i})$ by actuating the prismatic joints or the winch drives, permitting the control of the six degrees of freedom of the platform independently.

The feasible configurations of a hexapod can be defined implicitly, in a way similar to that followed in Sect. 3.5.2. Consider fixed and moving reference frames $\mathcal{F}_1$ and $\mathcal{F}_2$, respectively attached to the base and platform links, centred in O and P (Fig. 6.1). Let A_i and B_i be the anchor points of leg i at the base and the platform, $\mathbf{p}$ and $\mathbf{a}_i$ be the position vectors of P and A_i relative to $\mathcal{F}_1$, and $\mathbf{b}_i$ be the position vector of B_i relative to $\mathcal{F}_2$. We can represent any platform configuration by the pair $\mathbf{q} = (\mathbf{p}, \mathbf{R}) \in SE(3) = \mathbb{R}^3 \times SO(3)$, subject to the constraints

$$\mathbf{d}_i = \mathbf{p} + \mathbf{R}\,\mathbf{b}_i - \mathbf{a}_i, \tag{6.1}$$

$$d_i^2 = \mathbf{d}_i^{\mathsf{T}}\mathbf{d}_i, \tag{6.2}$$

$$\underline{d_i} < d_i < \overline{d_i}, \tag{6.3}$$

for $i = 1, \ldots, 6$, where $\mathbf{R}$ is the 3×3 rotation matrix that provides the orientation of $\mathcal{F}_2$ relative to $\mathcal{F}_1$. While Eqs. (6.1) and (6.2) make the leg lengths d_i explicit in terms of $\mathbf{p}$ and $\mathbf{R}$, the inequalities in (6.3) constrain such lengths to lie in $(\underline{d_i}, \overline{d_i})$.

As shown in Chap. 3, the orthonormality of $\mathbf{R}$ can be imposed in a variety of ways. Here, we simply assume that in Eq. (6.1) $\mathbf{R}$ is expressed as a function of $\boldsymbol{\tau}$, a tuple

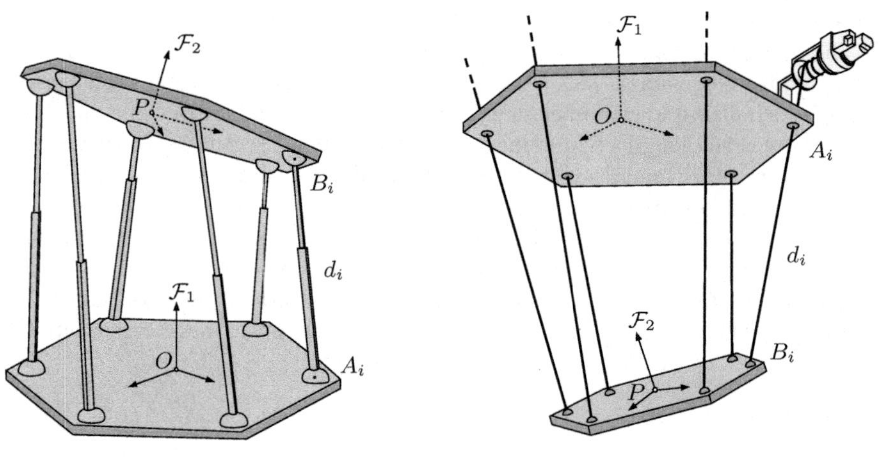

Fig. 6.1 Rigid-limbed (*left*) and cable-driven hexapods (*right*). A cable-driven hexapod is maintained under a stable position due to the action of gravity

of any three angles parameterising $SO(3)$, such as Euler angles, or tilt-and-torsion angles [4]. This allows for an easy formulation of planning problems in constant-angle slices of $SE(3)$, which are useful in parallel kinematic machines [4], and avoids the treatment of additional constraints needed in nonminimal representations of the rotation group. Although we then introduce representation singularities relative to the angles of choice [5, p. 31], this will not be problematic because the smoothness properties required to solve our planning problem will remain unaltered.

In practice, not all configurations satisfying Eqs. (6.1)–(6.3) are possible, because the leg forces must be kept within their admissible ranges. This is especially important in cable-driven robots, where all cables must be kept in tension to control the platform, but also in rigid-limbed ones to avoid excessive stress in the actuators or eventual breakage of the structure. For this reason we require any configuration to be *wrench-feasible*, in the sense that it must allow the platform to equilibrate any external wrench $\hat{\boldsymbol{w}}$ acting on it, subject to lie inside a prescribed six-dimensional region $\mathcal{K} \subset \mathbb{R}^6$.

The coordinates of $\hat{\boldsymbol{w}}$ are assumed to be given in the usual screw-theoretic form [6], i.e., with the first three components providing the net force on the platform, and the last three giving the net moment relative to O. The significance of $\mathcal{K}$, on the other hand, depends on the particular context of application. In payload transportation, for instance, $\mathcal{K}$ may be given by the gravitational wrench acting on the platform, and by slight perturbations introduced by inertia forces or external agents like the wind. In contact situations $\mathcal{K}$ might further depend on the contact wrench against the environment, which is in general subject to six-dimensional uncertainty.

In any case, the wrench-feasibility requirement for a given $\mathbf{q}$ implies that for each $\hat{\boldsymbol{w}} \in \mathcal{K}$ there must be a vector of admissible leg forces

$$\mathbf{f} = (f_1, \ldots, f_6) \in \mathcal{D} = (\underline{f_1}, \overline{f_1}) \times \cdots \times (\underline{f_6}, \overline{f_6})$$

satisfying

$$\mathbf{J} \cdot \mathbf{f} = \hat{\boldsymbol{w}},$$

where $(\underline{f_i}, \overline{f_i})$ is the range of force magnitudes that can be resisted by the ith leg, and $\mathbf{J}$ is the 6×6 screw Jacobian of the robot, which depends on $\mathbf{q}$ [7].

For ease of manipulation, we shall assume that $\mathcal{K}$ is a six-dimensional ellipsoid centred in a wrench $\hat{\boldsymbol{w}}_0 \in \mathbb{R}^6$, defined implicitly by the inequality

$$(\hat{\boldsymbol{w}} - \hat{\boldsymbol{w}}_0)^{\mathsf{T}} \mathbf{E} \, (\hat{\boldsymbol{w}} - \hat{\boldsymbol{w}}_0) \leq 1,$$

where $\mathbf{E}$ is a 6×6 positive-definite symmetric matrix.

For consistency, $\mathbf{J}$, $\hat{\boldsymbol{w}}_0$, and $\mathbf{E}$ have to be expressed in a same reference frame. Although any frame could be used, we shall assume for concreteness that $\mathbf{J}$, $\hat{\boldsymbol{w}}_0$, and $\mathbf{E}$ are all expressed in $\mathcal{F}_1$. In this frame, $\mathbf{J}$ takes the form

$$\mathbf{J} = \begin{bmatrix} \mathbf{u}_1 & \cdots & \mathbf{u}_6 \\ \mathbf{a}_1 \times \mathbf{u}_1 & \cdots & \mathbf{a}_6 \times \mathbf{u}_6 \end{bmatrix}$$

Fig. 6.2 From left to right and top to bottom: The Okuma PM-600 machining centre (*top*), and a close-up of the hexapod platform holding the tools (*bottom*) [8]; a worker programming and monitoring the TETRA RoboCrane from the National Institute of Standards and Technology (USA) [9], used to pick up and place a steel beam in a stanchion; a cable-driven hexapod for aircraft building and maintenance [9]; and a swell simulator [10, 11] [Photos courtesy of Okuma Corporation (Japan), the National Institute of Standards and Technology (USA), and Symétrie (France).]

where $\mathbf{u}_i = \mathbf{d}_i/d_i$, and both $\hat{\boldsymbol{w}}_0$ and $\mathbf{E}$ will be functions of $\mathbf{q}$ in general,

$$\hat{\boldsymbol{w}}_0 = \hat{\boldsymbol{w}}_0(\mathbf{q}),$$
$$\mathbf{E} = \mathbf{E}(\mathbf{q}),$$

depending on the task to which the hexapod is applied.

To have an idea, note that hexapod robots are currently used in a variety of contexts (Fig. 6.2), including machine tool technology, flight and automotive simulation, payload positioning, inspection, aircraft building and maintenance, swell simulation, oil-well fire fighting, or pipe assembling, to name a few, and each particular case demands a specific definition of $\hat{\boldsymbol{w}}_0$ and $\mathbf{E}$. A typical situation encountered in machine tools like those fabricated by the Okuma [8], Ingersoll [12], Toyoda [13], or Mikrolar companies [14], for example, is depicted in Fig. 6.3a. In this case, the legs must counteract the efforts given by the drilling wrench, $\hat{\boldsymbol{w}}_{drill}$, which can be bound by a six-dimensional ellipsoid centred in some point $\hat{\boldsymbol{w}}_0$, with $\mathbf{E}$ accounting for the uncertainty of the operation. Note that if the weight of the platform-tool, $\hat{\boldsymbol{w}}_{weight}$, can be neglected, then $\hat{\boldsymbol{w}}_0 = \hat{\boldsymbol{w}}_{drill}$ is constant in $\mathcal{F}_1$, but that it must be

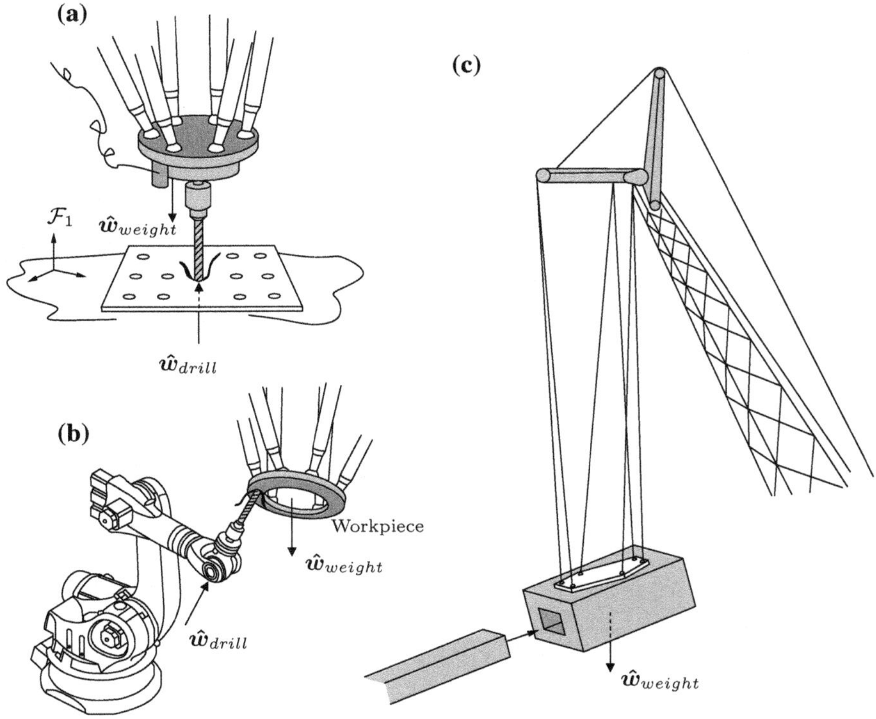

Fig. 6.3 Hexapods working in different scenarios

shifted by $\hat{\boldsymbol{w}}_{weight}$ otherwise, which may depend on $\mathbf{q}$ in general. A more complex drilling operation could be performed if the workpiece moves attached to the hexapod and, at the same time, is being drilled in a coordinate manner by a tool driven by another robot (Fig. 6.3b). In such a situation it would be simpler to initially express $\hat{\boldsymbol{w}}_0$ and $\mathbf{E}$ in the platform frame, where $\mathbf{E}$ is constant and $\hat{\boldsymbol{w}}_0$ is either constant or shifted by $\hat{\boldsymbol{w}}_{weight}$ as before, but $\hat{\boldsymbol{w}}_0$ and $\mathbf{E}$ will finally depend on $\mathbf{q}$ when expressed in $\mathcal{F}_1$. In applications involving a high payload that must be manipulated, like the one depicted in Fig. 6.3c, it might be interesting to consider slight perturbations of the weight introduced by inertia forces or external agents like the wind. Since the weight always points in the same direction, $\mathbf{E}$ and $\hat{\boldsymbol{w}}_0$ can initially be defined in a frame that translates with the platform, with its origin at the centre of mass, where $\mathbf{E}$ and $\hat{\boldsymbol{w}}_0 = \hat{\boldsymbol{w}}_{weight}$ are constant. If we then express $\mathbf{E}$ and $\hat{\boldsymbol{w}}_0$ in $\mathcal{F}_1$, their expressions will be dependent on $\mathbf{q}$ again. Section 6.5 provides further details on how to obtain $\mathbf{E}(\mathbf{q})$ and $\hat{\boldsymbol{w}}_0(\mathbf{q})$ in common scenarios.

6.2 The Planning Problem

Let us now define the wrench-feasible C-space of the mechanism, $\mathcal{C}$, as the set of configurations $\mathbf{q} \in SE(3)$ that are wrench-feasible, and also satisfy Eqs. (6.1)–(6.3) for $i = 1, \ldots, 6$. The planning problem we confront, thus, boils down to computing a *path* joining two given configurations of $\mathcal{C}$, $\mathbf{q}_s$ and $\mathbf{q}_g$, i.e., a continuous map

$$\kappa : [0, 1] \longrightarrow \mathcal{C}$$

such that $\kappa(0) = \mathbf{q}_s$ and $\kappa(1) = \mathbf{q}_g$. To tackle this problem, we next construct a smooth manifold suitable to navigate $\mathcal{C}$ by means of the numerical continuation method of Chap. 5.

6.2.1 A Characterisation of the Wrench-Feasible C-Space

For a given configuration $\mathbf{q}$, let $\mathbf{f}_0$ be the vector of leg forces corresponding to the centre wrench of the ellipsoid, $\hat{\boldsymbol{w}}_0$, which satisfies

$$\mathbf{J} \cdot \mathbf{f}_0 = \hat{\boldsymbol{w}}_0. \tag{6.4}$$

By noting that

$$\mathbf{J}(\mathbf{f} - \mathbf{f}_0) = \hat{\boldsymbol{w}} - \hat{\boldsymbol{w}}_0,$$

it is easy to see that the set $\mathcal{F}$ of leg forces $\mathbf{f}$ corresponding to all wrenches $\hat{\boldsymbol{w}} \in \mathcal{K}$ is also an ellipsoid, which is given by

$$(\mathbf{f} - \mathbf{f}_0)^\mathsf{T} \mathbf{B}\, (\mathbf{f} - \mathbf{f}_0) \leq 1,$$

where $\mathbf{B} = \mathbf{J}^\mathsf{T} \mathbf{E}\, \mathbf{J}$. This ellipsoid may be bounded in all directions or unbounded in some, depending on whether $det(\mathbf{J}) \neq 0$ or not. However, Sect. 6.3.1 shows that $\mathbf{J}$ is nonsingular for all $\mathbf{q} \in \mathcal{C}$, so that $\mathcal{F}$ will always be a bounded ellipsoid in our case (Fig. 6.4). It is worth to realise here that, since $\mathbf{J}$ is full rank for all $\mathbf{q} \in \mathcal{C}$, the navigation of $\mathcal{C}$ implicitly avoids the forward singularity locus of the platform. As a result, the control issues due to the indeterminacy of the output velocity, or loss of rigidity in such locus, will not be encountered during the execution of the planned path.

Now, for $\mathbf{q}$ to be wrench-feasible, we must have $\mathcal{F} \subseteq \mathcal{D}$, which can be checked as follows. Consider the vector $\mathbf{v}_i$ that gives the offset from the centre of $\mathcal{F}$ to the point of $\mathcal{F}$ attaining the largest f_i value, $\mathbf{f}_0 + \mathbf{v}_i$ (Fig. 6.4). Symmetrically, $\mathbf{f}_0 - \mathbf{v}_i$ is the point of $\mathcal{F}$ with the smallest f_i value. Using Lagrange multipliers, one can see that $\mathbf{v}_i$ is the *unique* vector satisfying

$$\mathbf{v}_i^\mathsf{T} \mathbf{B}\, \mathbf{v}_i = 1, \tag{6.5}$$

$$\mathbf{B}^i \mathbf{v}_i = \mathbf{0}, \tag{6.6}$$

$$v_{i,i} > 0, \tag{6.7}$$

where $\mathbf{B}^i$ denotes the matrix $\mathbf{B}$ with its ith row removed, and $v_{i,i}$ is the i-th component of $\mathbf{v}_i$. Observe that if $\mathbf{J}$ is nonsingular, then both $\mathbf{B}$ and $\mathbf{B}^i$ are full row rank, and certainly there is exactly one vector $\mathbf{v}_i$ satisfying Eqs. (6.5)–(6.7). Using this vector, we can say that if $det(\mathbf{J}) \neq 0$, then $\mathcal{F}$ is contained in $\mathcal{D}$ whenever

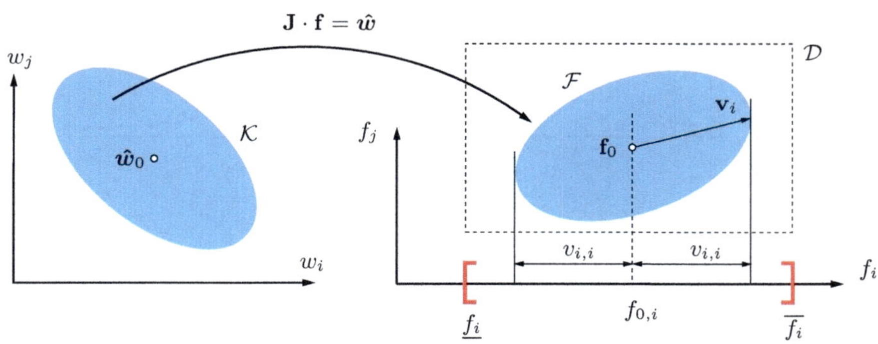

Fig. 6.4 The map $\mathbf{J} \cdot \mathbf{f} = \hat{\boldsymbol{w}}$ is used to transform the wrench ellipsoid $\mathcal{K}$ into the leg force ellipsoid $\mathcal{F}$. The vector $\mathbf{v}_i$ provides the maximum and minimum values of f_i within $\mathcal{F}$

$$f_{0,i} - v_{i,i} > \underline{f_i}, \tag{6.8}$$

$$f_{0,i} + v_{i,i} < \overline{f_i}, \tag{6.9}$$

for $i = 1, \ldots, 6$. As a result, $\mathcal{C}$ can be characterised as the set of points $\mathbf{q} = (\mathbf{p}, \mathbf{R}) \in SE(3)$ satisfying Eqs. (6.1)–(6.9) for some value of the variables $\mathbf{d}_i$, d_i, $\mathbf{f}_0$ and $\mathbf{v}_i$.

6.2.2 Conversion into Equality Form

As done in Sect. 5.2, we now need to convert Eqs. (6.3), and (6.7)–(6.9) into equality form to be able to apply the numerical continuation method. The conversion can be achieved by noting that we can replace Eq. (6.3) by the two equivalent conditions

$$(d_i - \underline{d_i}) \cdot (\overline{d_i} - d_i) \cdot g_i = 1, \tag{6.10}$$
$$g_i > 0,$$

where g_i is a newly-defined auxiliary variable (Fig. 6.5a). In apparence, we have not skipped the use of inequalities with this change, but from the graph of Fig. 6.5a we see that if a configuration $\mathbf{q}$ corresponds to a value $d_i = a \in (\underline{d_i}, \overline{d_i})$, then any other configuration found from $\mathbf{q}$ by continuation subject to Eq. (6.10) will always satisfy $\underline{d_i} < d_i < \overline{d_i}$. In other words, the constraint $g_i > 0$ can be neglected under such a continuation scheme.

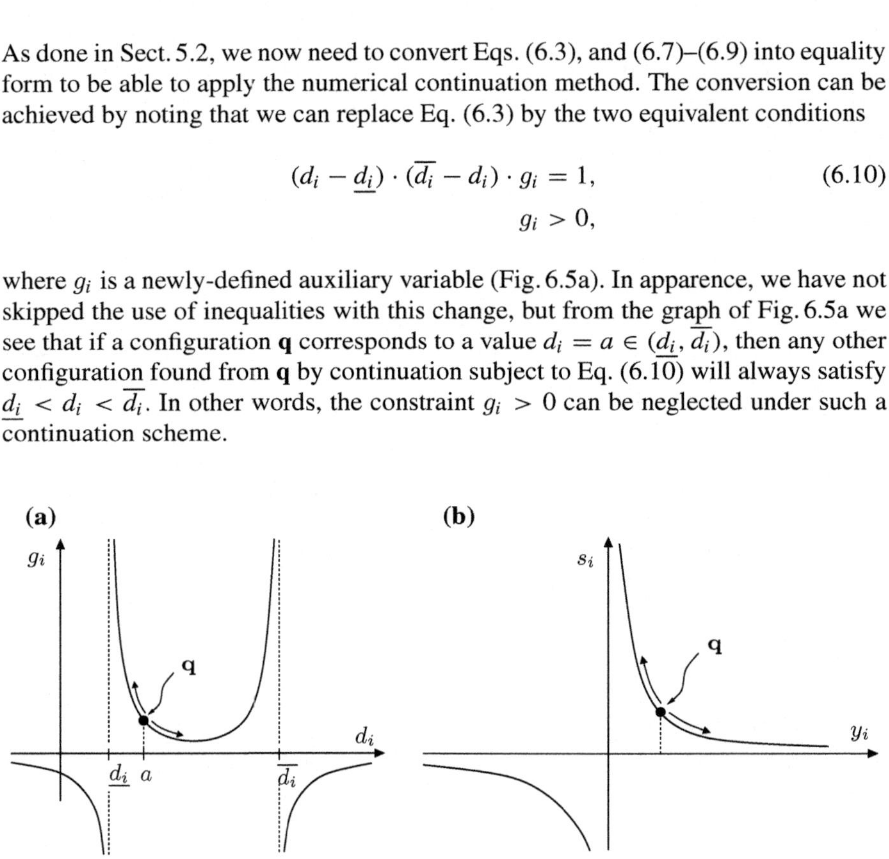

Fig. 6.5 **a** The graph of Eq. (6.10) shows that if we explore the solution set of Eq. (6.10) by continuation from some $\mathbf{q} \in \mathcal{C}$ corresponding to $d_i = a$, the constraints $\underline{d_i} < d_i < \overline{d_i}$ will always be satisfied. **b** The graph of $y_i s_i = 1$, where $y_i = f_{0,i} - v_{i,i} - \underline{f_i}$, shows that the same applies to Eq. (6.11)

Similarly, Eqs. (6.8) and (6.9) can be replaced by

$$(f_{0,i} - v_{i,i} - \underline{f_i}) \cdot s_i = 1, \tag{6.11}$$
$$(\overline{f_i} - f_{0,i} - v_{i,i}) \cdot t_i = 1, \tag{6.12}$$
$$s_i > 0, \quad t_i > 0$$

where s_i and t_i play a role analogous to that of g_i in Eq. (6.10). From the graph in Fig. 6.5b, for example, it is clear that the quantity

$$y_i = f_{0,i} - v_{i,i} - \underline{f_i}$$

will remain positive, and hence $f_{0,i} - v_{i,i} > \underline{f_i}$ when marching continuously from a given $\mathbf{q}$ with $y_i > 0$. The same argument applies to Eq. (6.12), which allows us to replace Eqs. (6.8) and (6.9) by Eqs. (6.11) and (6.12), neglecting the constraints $s_i > 0$ and $t_i > 0$ during the continuation-based search.

Finally, Eq. (6.7) can be directly neglected because $v_{i,i} \neq 0$ for all i in any vector satisfying Eqs. (6.5) and (6.6). Certainly, observe that $\mathbf{B}\mathbf{v}_i$ is all zeros except in its i-th component due to Eq. (6.6). If it were $v_{i,i} = 0$ for some i, this would imply $\mathbf{v}_i^\mathsf{T}\mathbf{B}\mathbf{v}_i = 0$, contradicting Eq. (6.5). Therefore, if our continuation method starts from a value of $\mathbf{v}_i$ with $v_{i,i} > 0$, and it is compliant with Eqs. (6.5) and (6.6), Eq. (6.7) will be naturally fulfilled.

6.2.3 The Navigation Manifold

The system formed by Eqs. (6.1), (6.2), (6.4)–(6.6), and (6.10)–(6.12) can be compactly written as

$$\mathbf{F}(\mathbf{x}) = \mathbf{0}, \tag{6.13}$$

where $\mathbf{x}$ is a tuple encompassing all of its variables

$$\mathbf{p}, \boldsymbol{\tau}, \mathbf{d}_i, d_i, g_i, \mathbf{f}_0, \mathbf{v}_i, s_i, t_i,$$

for $i = 1, \ldots, 6$. The solution set of this system,

$$\mathcal{M} = \{\mathbf{x} : \mathbf{F}(\mathbf{x}) = \mathbf{0}\},$$

will be called the *navigation* manifold hereafter, because it has the necessary properties to connect $\mathbf{q}_s$ and $\mathbf{q}_g$ by numerical continuation.

To see this point, consider the subset of $\mathcal{M}$

$$\mathcal{M}^+ = \{\mathbf{x} \in \mathcal{M} : g_i > 0, s_i > 0, t_i > 0, v_{i,i} > 0\},$$

Clearly, any continuous path in $\mathcal{C}$ is represented by a continuous path in $\mathcal{M}^+$, and *vice versa*. Thus, the original problem of computing a path of $\mathcal{C}$ connecting $\mathbf{q}_s$ and $\mathbf{q}_g$ boils down to computing a path in $\mathcal{M}^+$ connecting two points of $\mathcal{M}^+$, $\mathbf{x}_s$ and $\mathbf{x}_g$, corresponding to $\mathbf{q}_s$ and $\mathbf{q}_g$. However, since g_i, s_i, t_i, and $v_{i,i}$ never vanish in $\mathcal{M}$ (Sect. 6.2.2), $\mathcal{M}^+$ and its complement $\mathcal{M} \setminus \mathcal{M}^+$ are disconnected, and when marching from $\mathbf{x}_s$ to $\mathbf{x}_g$ by continuation in $\mathcal{M}$ we shall be moving within $\mathcal{M}^+$ actually. Hereafter, therefore, we shall only need $\mathcal{M}$ and Eq. (6.13) in our algorithms.

$\mathcal{M}$ is six-dimensional and highly nonlinear, but in Sect. 6.3 we prove that it is smooth everywhere, so that every point $\mathbf{x}$ has a well-defined tangent space $\mathcal{T}_\mathbf{x}\mathcal{M}$. As noted in the previous chapter, this greatly simplifies the application of a continuation method to connect $\mathbf{x}_s$ and $\mathbf{x}_g$, because no bifurcations, sharpnesses, or dimension changes will be found in $\mathcal{M}$, avoiding the need of elaborate branch-switching procedures [15]. To connect $\mathbf{x}_s$ to $\mathbf{x}_g$, therefore, we shall use the same method proposed in Sect. 5.3.

6.2.4 Addition of Pose Constraints

It is worth adding that in many applications (such as in painting, polishing, or cleaning of ship hulls, wings, or building façades) the platform is further confined to move within a lower-dimensional subset of $\mathcal{C}$ defined by geometric or contact constraints on its pose. As exemplified in Sect. 6.4.2, we can directly add such constraints to Eq. (6.13) if we wish, written either in the parametric form

$$\begin{bmatrix} \mathbf{p} \\ \boldsymbol{\tau} \end{bmatrix} = \boldsymbol{\Omega}(\boldsymbol{\lambda}), \tag{6.14}$$

where $\boldsymbol{\Omega}$ is an arbitrary smooth function of any set of parameters $\boldsymbol{\lambda}$, or in the implicit form

$$\mathbf{C}(\mathbf{p}, \boldsymbol{\tau}) = \mathbf{0}, \tag{6.15}$$

where $\mathbf{C}(\mathbf{p}, \boldsymbol{\tau})$ is any smooth function with a full-rank Jacobian $\mathbf{C}_{\mathbf{p},\boldsymbol{\tau}}$. Section 6.3.3 shows that, again, the resulting system of equations yields a smooth manifold suitable to the continuation strategy of Chap. 5.

6.3 Proofs of the Properties

Before illustrating the planning strategy with a few examples, we next prove three properties that have been assumed so far. Namely, that the screw Jacobian $\mathbf{J}$ is nonsingular at all points of $\mathcal{M}$, that $\mathcal{M}$ is smooth, and that its lower-dimensional subsets defined by Eqs. (6.14) and (6.15) are also smooth.

6.3.1 Nonsingularity of the Screw Jacobian

It is easy to prove that $\mathbf{J}$ is full rank for all $\mathbf{q} \in \mathcal{C}$ by contradiction. Let us assume that there were a configuration $\mathbf{q} \in \mathcal{C}$ for which $\mathbf{J}$ were rank deficient. Then, since $\mathbf{B}$ would be rank deficient as well, we would have $\dim(\ker \mathbf{B}) \geq 1$. In this case, there would be some i for which $\ker \mathbf{B} = \ker \mathbf{B}^i$, i.e. all solutions $\mathbf{v}_i$ of $\mathbf{B}^i \mathbf{v}_i = \mathbf{0}$ would lie in $\ker \mathbf{B}$. This would imply that $\mathbf{v}_i^{\mathsf{T}} \mathbf{B} \mathbf{v}_i = 0$ for such i, which contradicts Eq. (6.5). Therefore, $\mathbf{J}$ must be nonsingular for all $\mathbf{q} \in \mathcal{C}$.

6.3.2 Smoothness of the Navigation Manifold

To prove that $\mathcal{M}$ is a smooth manifold, it is sufficient to show that the map $\mathbf{F}(\mathbf{x})$ is differentiable and has a full rank Jacobian $\mathbf{F}_{\mathbf{x}}$ for all $\mathbf{x} \in \mathcal{M}$. Then, the smoothness of $\mathcal{M}$ follows from the implicit function theorem [16].

By construction, all functions intervening in $\mathbf{F}(\mathbf{x})$ are clearly differentiable. After reordering the equations, $\mathbf{F}_{\mathbf{x}}$ can be expressed in the following block-triangular form

$$
\mathbf{F}_{\mathbf{x}} = \left[\begin{array}{c|c|c|c|c}
\boldsymbol{\Phi}_{\mathbf{y}} & & & & \\
\hline
* & \mathbf{J} & & & \\
\hline
* & & \begin{array}{ccc} \begin{array}{c} 2\mathbf{v}_1^{\mathsf{T}}\mathbf{B} \\ \mathbf{B}^1 \end{array} & & \\ & \ddots & \\ & & \begin{array}{c} 2\mathbf{v}_6^{\mathsf{T}}\mathbf{B} \\ \mathbf{B}^6 \end{array} \end{array} & & \\
\hline
 & * & * & \mathbf{S} & \\
\hline
 & * & * & & \mathbf{T}
\end{array}\right],
$$

where the empty blocks represent zero-matrices and the asterisks indicate nonzero blocks. The block $\boldsymbol{\Phi}_{\mathbf{y}}$ is of size 30×36, and the remaining diagonal blocks are all of size 6×6.

To see that $\mathbf{F}_\mathbf{x}$ is full rank, we next check that all blocks in its diagonal are full-rank:

1. The block $\boldsymbol{\Phi}_\mathbf{y}$ is the Jacobian of the subsystem $\boldsymbol{\Phi}(\mathbf{y}) = \mathbf{0}$ formed by Eqs. (6.1), (6.2), and (6.10), where

$$\mathbf{y} = (\mathbf{p}, \boldsymbol{\tau}, \mathbf{d}_1, \ldots, \mathbf{d}_6, d_1, \ldots, d_6, g_1, \ldots, g_6). \tag{6.16}$$

   This Jacobian takes the form

$$\boldsymbol{\Phi}_\mathbf{y} = \left[\begin{array}{c|c|c|c|c} * & * & -\mathbf{I}_{18\times 18} & & \\ \hline & & * & \mathbf{L} & \\ \hline & & & * & \mathbf{G} \end{array}\right], \tag{6.17}$$

   where the block columns correspond to the derivatives of the variables in $\mathbf{y}$ in the order of Eq. (6.16), and $\mathbf{L}$ and $\mathbf{G}$ are 6×6 diagonal matrices with $-2d_i$ and $(d_i - \underline{d_i}) \cdot (\overline{d_i} - d_i)$ in their entries, respectively. Observe that $\boldsymbol{\Phi}_\mathbf{y}$ is full rank, because the entries in $\mathbf{L}$ and $\mathbf{G}$ do not vanish over $\mathcal{M}$, by virtue of Eq. (6.10) and the fact that $\underline{d_i} > 0$.
2. Sect. 6.3.1 proves that the block $\mathbf{J}$ is full rank for all $\mathbf{q} \in \mathcal{C}$, and the same argument shows that it must be so for all $\mathbf{x} \in \mathcal{M}$.
3. The 6×6 block matrices involving $\mathbf{B}$ and $\mathbf{B}^i$ can only be rank deficient if $v_{i,i} = 0$ but, as explained in Sect. 6.2.2, this can never happen in $\mathcal{M}$.
4. Finally, $\mathbf{S}$ and $\mathbf{T}$ are 6×6 diagonal matrices whose elements are the terms $f_{0,i} - v_{i,i} - \underline{f_i}$ and $\overline{f_i} - f_{0,i} - v_{i,i}$, respectively, which do not vanish because of Eqs. (6.11) and (6.12).

6.3.3 Smoothness of Lower-Dimensional Subsets

The subsets of $\mathcal{M}$ defined by adding Eq. (6.14) or Eqs. (6.15)–(6.13) are also smooth manifolds. When adding Eq. (6.14), the Jacobian of the resulting system of equations takes the form

$$\left[\begin{array}{c|c|c|c} \Omega_\lambda & -\mathbf{I}_{6\times 6} & & \\ \hline & \boldsymbol{\Phi}_1 & \boldsymbol{\Phi}_2 & \\ \hline & * & * & \mathbf{P} \end{array}\right], \tag{6.18}$$

where $\boldsymbol{\Phi}_1$ and $\boldsymbol{\Phi}_2$ are the matrices formed by the first two and last three block-columns of $\boldsymbol{\Phi}_\mathbf{y}$ in Eq. (6.17), respectively, and $\mathbf{P}$ is the submatrix of $\mathbf{F}_\mathbf{x}$ obtained by removing its first block-row and block-column. The Jacobian in (6.18) is full rank because the square blocks $-\mathbf{I}_{6\times 6}$, $\boldsymbol{\Phi}_2$ and $\mathbf{P}$ are all full rank. Similarly, if we add Eqs. (6.15)–(6.13) the resulting Jacobian is

$$\left[\begin{array}{c|c|c} \mathbf{C}_{\mathbf{p},\mathcal{T}} & & \\ \hline \boldsymbol{\Phi}_1 & \boldsymbol{\Phi}_2 & \\ \hline * & * & \mathbf{P} \end{array}\right],$$

which is clearly full rank because $\mathbf{C}_{\mathbf{p},\mathcal{T}}$ is nonsingular according to the assumptions in Sect. 6.2.4.

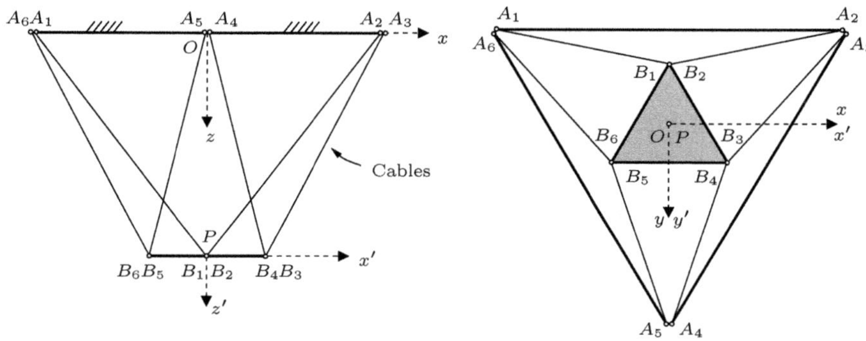

Robot	Base	Platform
1	$A_1 = (-200, -115.47, 0)$	$B_1 = (0, -115.47, 0)$
	$A_2 = (200, -115.47, 0)$	$B_2 = (0, -115.47, 0)$
	$A_3 = (200, -115.47, 0)$	$B_3 = (100, 57.74, 0)$
	$A_4 = (0, 230.94, 0)$	$B_4 = (100, 57.74, 0)$
	$A_5 = (0, 230.94, 0)$	$B_5 = (-100, 57.74, 0)$
	$A_6 = (-200, -115.47, 0)$	$B_6 = (-100, 57.74, 0)$
2	$A_1 = (-231.62, -136.18, 0)$	$B_1 = (0, -89.15, 0)$
	$A_2 = (231.62, -136.18, 0)$	$B_2 = (0, -89.15, 0)$
	$A_3 = (233.74, -132.50, 0)$	$B_3 = (77.21, 44.57, 0)$
	$A_4 = (2.13, 268.67, 0)$	$B_4 = (77.21, 44.57, 0)$
	$A_5 = (-2.13, 268.67, 0)$	$B_5 = (-77.21, 44.57, 0)$
	$A_6 = (-233.74, -132.50, 0)$	$B_6 = (-77.21, 44.57, 0)$

Fig. 6.6 Geometry of the two cable-driven robots considered. (*Top*) Front and *top* views of their octahedral architecture in a reference configuration. (*Bottom*) Coordinates of A_i and B_i in mm, expressed in $\mathcal{F}_1 = Oxyz$ and $\mathcal{F}_2 = Px'y'z'$ respectively

6.4 Case Studies

We next show the performance of the method in a few test cases. Initially, we apply the planner to the two cable-driven robots described in Fig. 6.6, which are inspired in the NIST RoboCrane [17]. In Sect. 6.4.1, we use the planner to compute paths for "Robot 1" in two-dimensional slices of $\mathcal{C}$ obtained by fixing four pose parameters. This shows how complex the wrench-feasible C-space can be even in simple cases, and stresses the advantages of our planning approach in comparison to previous methods that only guarantee path feasibility at discrete points [18–20]. In Sect. 6.4.2, we then apply the planner to "Robot 2", which corresponds to a real prototype called "Hexacrane" constructed in the authors' lab. This robot is used to illustrate motions subject to nontrivial geometric constraints, and also free-flying motions in the full six-dimensional space. Finally, in Sect. 6.4.3 we apply the method to the Gough-Stewart platform studied in Li et al. [21], and to the INRIA Left Hand hexapod [22].

All results have been obtained with an implementation in C of the method which is available in [23], executed in a MacBook Pro computer equipped with a 2.66 GHz Intel Core i7 processor.

Because of its attractive properties in parallel machines, the implementation adopts the tilt-and-torsion parameterisation of $SO(3)$, for which

$$\mathbf{R} = \mathbf{R}_z(\phi)\mathbf{R}_y(\theta)\mathbf{R}_z(\sigma - \phi),$$

where ϕ, θ, and σ are the azimuth, tilt, and torsion angles respectively [4]. Thus, $\tau = (\phi, \theta, \sigma)$ in this section, and the algorithms take into account that the angular coordinates differing in multiples of 2π refer to the same angle. Note also that, although some of the tested robots follow the octahedral architecture, the planner remains applicable to general hexapods, with cable anchor points not necessarily coincident in pairs. Computation times for all test cases are summarised in Sect. 6.4.4.

6.4.1 Planning in Illustrative Slices

In this example, Robot 1 is required to withstand a load of 1 Newton applied at a point P_m with position vector $\mathbf{p}_m = (30, 14, -21)$ mm in frame $\mathcal{F}_2$. Note that the weight of this load corresponds to a constant wrench $\hat{\boldsymbol{w}}_0 = (0, 0, 1, 0, 0, 0)$ (in SI units) if expressed in a frame $\mathcal{F}_3$ defined parallel to $\mathcal{F}_1$ and translating with P_m. The bounded perturbations of this wrench will be represented by the ellipsoid $\mathcal{K}$ centred in $\hat{\boldsymbol{w}}_0$ with $\mathbf{E} = 10^4\mathbf{I}_{6\times 6}$ in SI units, also expressed in $\mathcal{F}_3$. Both $\hat{\boldsymbol{w}}_0$ and $\mathbf{E}$ can be expressed in $\mathcal{F}_1$ using Eqs. (6.21) and (6.22). The tensions and lengths for all cables are constrained to the ranges $f_i \in (0.05, 0.5)$ N and $d_i \in (100, 550)$ mm.

Figure 6.7 shows several slices of the wrench-feasible C-space of the robot, computed in Matlab using dense discretisation. Whereas the left column shows slices in which P and σ are held fixed, the right column shows others in which the full

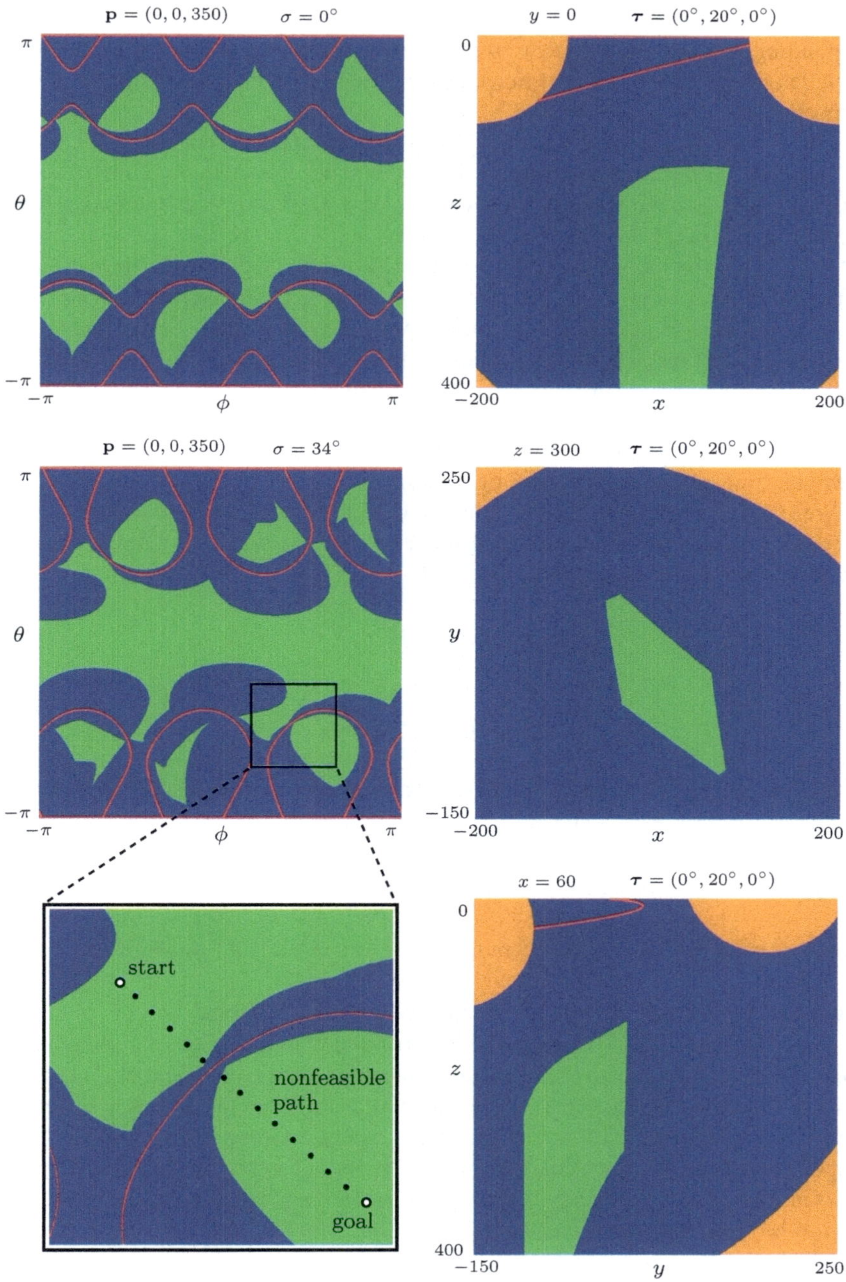

Fig. 6.7 Slices of $\mathcal{C}$ for Robot 1, obtained by dense discretisation

orientation $\boldsymbol{\tau}$ and one of the coordinates of P are constant. The configurations corresponding to $\mathcal{C}$ are indicated in green, while those that cannot be reached due to cable lengths or tensions out of range are represented by the orange and blue areas, respectively.

The symmetries in the slices of the top row appear because (ϕ, θ, σ) and $(\phi + \pi, -\theta, \sigma)$ represent the same orientation under the chosen parameterisation. To avoid this double coverage of $SO(3)$, we only need to restrict the expansion of the atlas to the range $\theta \in [0, \pi]$.

Figure 6.7 also shows, in red, the singularity curves where $det(\mathbf{J}(\mathbf{q})) = 0$, computed under no constraints on the cable tensions using the method of Chap. 3. It can be observed that, as expected, $\mathcal{C}$ naturally avoids crossing such curves, and that the navigation between two configurations is not trivial because $\mathcal{C}$ is in general non-convex and may have very close connected components. The bottom-left plot of Fig. 6.7 exemplifies how, in particular, evaluating the wrench-feasibility conditions at discrete points, as done e.g. in [18–20], could result in erroneous paths linking configurations of $\mathcal{C}$ separated by singularity curves.

We apply our method to resolve a planning query in the left-centre slice of Fig. 6.7, hence exploring a manifold of dimension $n = 2$. Since here the platform can only rotate, we define the functions of Eqs. (5.13) and (5.14) as

$$C_{ST}(\mathbf{x}_j, \mathbf{x}_k) = \|log(\mathbf{R}(\boldsymbol{\tau}_j)^{\mathsf{T}}\mathbf{R}(\boldsymbol{\tau}_k))\|,$$

and

$$H_{ST}(\mathbf{x}_i) = c(\mathbf{x}_i, \mathbf{x}_g).$$

to implement the A* and Gready Best-First strategies of the planner. For two orientations given by $\mathbf{R}(\boldsymbol{\tau}_j)$ and $\mathbf{R}(\boldsymbol{\tau}_k)$, $C_{ST}(\mathbf{x}_j, \mathbf{x}_k)$ gives the angle of the axis-angle representation of $\mathbf{R}(\boldsymbol{\tau}_j)^{\mathsf{T}}\mathbf{R}(\boldsymbol{\tau}_j)$, which is a standard metric of $SO(3)$ [24]. The start and goal configurations $\mathbf{q}_s$ and $\mathbf{q}_g$ are given by the $\boldsymbol{\tau}$ values

$$\boldsymbol{\tau}_s = (-2.3, 1.6, 0.59) \text{ rad}, \qquad \boldsymbol{\tau}_g = (1.7, 1.7, 0.59) \text{ rad},$$

and by the position of P, which is fixed at $\mathbf{p} = (0, 0, 350)$ mm.

Figure 6.8 shows the path returned by the planner, in red, after smoothing it. The green mesh corresponds to the full atlas of the connected component of $\mathcal{C}$ attainable from $\mathbf{q}_s$, and the shaded area indicates the part of this component that was actually explored to obtain the red path using the A* strategy. Green and blue shading correspond to closed and open charts respectively, and it can be seen that the algorithm takes into account the topology of the angle variables, allowing the planner to traverse from $\phi = -\pi$ to $\phi = \pi$, or *vice versa*.

Observe that a naive approach based on interpolating the start and goal configurations would violate the constraints of $\mathcal{C}$, giving rise to uncontrollable motions or unaffordable cable tensions. The planned path, in contrast, correctly avoids these situations and guarantees control of the platform at all points, while keeping cable

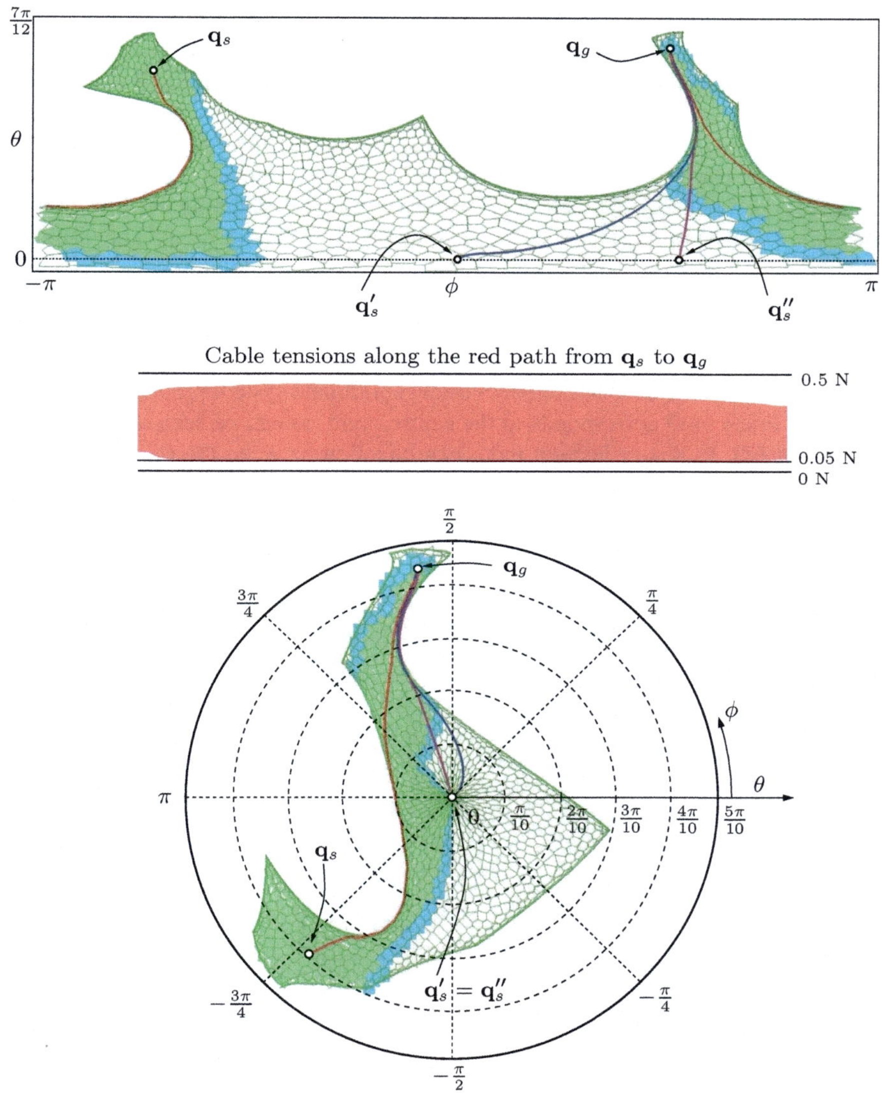

Fig. 6.8 The Cartesian (*top*) and cylindrical coordinate (*bottom*) plots of the three planning queries resolved in Sect. 6.4.1. The *green* mesh of hexagon-like polygons corresponds to the polytopes $\mathcal{P}_i$ of the whole atlas obtained from $\mathbf{q}_s$. Only part of such atlas is generated when trying to connect $\mathbf{q}_s$ to $\mathbf{q}_g$ under the A* strategy, shown shaded in *green* (closed charts) and *blue* (open charts). The configurations $\mathbf{q}'_s$ and $\mathbf{q}''_s$ are in fact the same one, because they lie on the $\theta = 0$ line. The plot in the centre of the figure shows the envelope of cable tensions along the *red* path from $\mathbf{q}_s$ to $\mathbf{q}_g$. As demanded, the tensions stay in the range (0.05, 0.5) N

lengths and tensions within their limits. The snippet in Fig. 6.8 also shows that the envelope of cable forces is admissible along the movement. Furthermore, because we rely on continuation, the synthesised path will not misleadingly bridge two disjoint components of $\mathcal{C}$, a property that cannot be ensured by discretisation-based strategies.

The line $\theta = 0$ of the Cartesian plot in Fig. 6.8 is known to be a representation singularity because all of its points correspond to a same orientation [4]. To illustrate that this is not a problem in practice, we show the results of planning two movements starting at distinct points of the $\theta = 0$ line, $\mathbf{q}_s'$ and $\mathbf{q}_s''$, both leading to $\mathbf{q}_g$. The resulting paths (shown in blue and purple) are different because the nature of the algorithm does not capture the fact that there is no cost of moving between two configurations with $\theta = 0$. However, $\mathcal{M}$ is smooth despite the singularity, and the planner has no problem in computing feasible movements in the two cases. An alternative way to represent these results is to use the cylindrical coordinate system proposed by [4], in which the line $\theta = 0$ corresponds to the centre point, as can be seen in the bottom plot of Fig. 6.8. In this projection there is a bijection between the (ϕ, θ) pairs and their corresponding orientations, and we can see that the blue and purple paths are similar, only differing in the vicinity of $\theta = 0$.

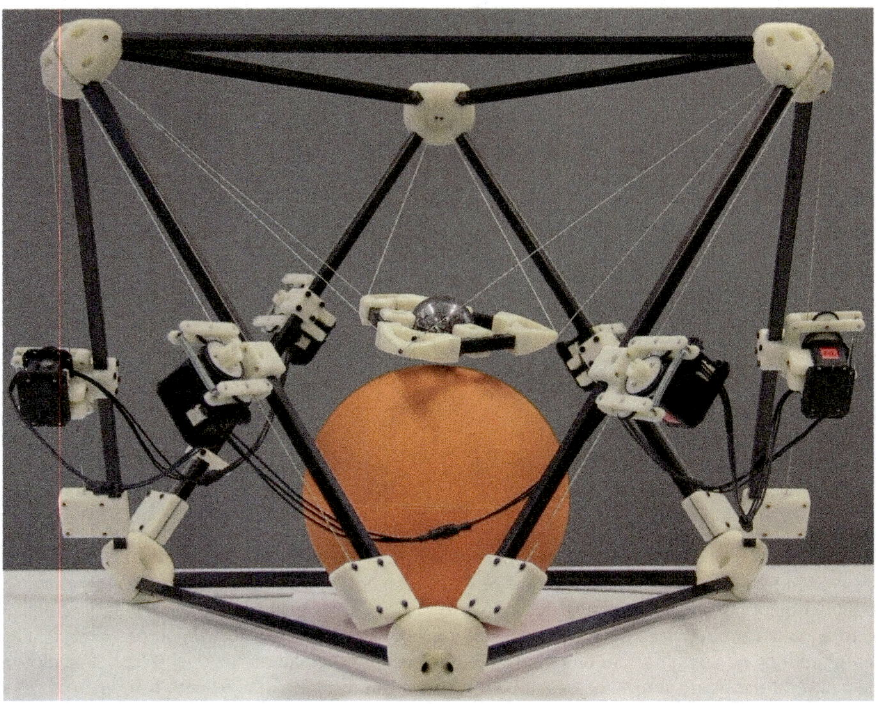

Fig. 6.9 The IRI Hexacrane, a cable-driven hexapod constructed at Institut de Robòtica i Informàtica Industrial (IRI)

6.4.2 Planning in the IRI Hexacrane

In order to mimic a situation in which the platform is subject to geometric constraints, we next apply our approach to the robot of Fig. 6.9, which is meant to perform insertion tasks on the surface of a sphere. For operation purposes, the platform is required to move tangentially to the sphere with zero torsion. Using Eq. (6.14), these conditions can be written as follows:

$$\left.\begin{aligned} \mathbf{p} &= \mathbf{r}_c + r_s \begin{bmatrix} \cos\alpha_2 \cos\alpha_1 \\ \cos\alpha_2 \sin\alpha_1 \\ -\sin\alpha_2 \end{bmatrix} \\ \phi &= \alpha_1 - \pi \\ \theta &= \tfrac{\pi}{2} - \alpha_2 \\ \sigma &= 0 \end{aligned}\right\}, \tag{6.19}$$

where $\mathbf{r}_c = (x_s, y_s, z_s)$ and r_s indicate the sphere centre and radius, and α_1 and α_2 are two angular parameters. Thus, $\boldsymbol{\lambda} = (\alpha_1, \alpha_2)$ in this case, and the navigation manifold $\mathcal{M}$ is of dimension $n = 2$ after adding Eqs. (6.19)–(6.13).

The points A_i and B_i correspond to those of Robot 2 in Fig. 6.6, and the sphere is of radius 100 mm, with its centre located at $\mathbf{r}_c = (0, 0, 306)$ mm in frame $\mathcal{F}_1$. However, since a small distance between the platform and the sphere needs to be kept, a value of $r_s = 130$ mm is used in Eq. (6.19). The platform weight is of 0.6 Kg, with its centre of mass located in P, and we use the same matrix $\mathbf{E} = 10^4\, \mathbf{I}_{6\times6}$ as before. Cable tensions are limited to $f_i \in (0.1, 6.58)$ N, and the feasible lengths must satisfy $d_i \in (200, 600)$ mm, for $i = 1, \ldots, 6$.

The resulting C-space is shown in Fig. 6.10 projected on the sphere, using the same drawing conventions of Fig. 6.8. The initial configuration, and the configurations where the insertion tasks are to be done are referred to as $\mathbf{q}_1$, $\mathbf{q}_2$, and $\mathbf{q}_3$, respectively, and correspond to the $\boldsymbol{\lambda}$ values $(0.55, \frac{\pi}{2})$, $(0.55, 0.75)$, and $(2.63, 0.75)$.

If we ask the planner to synthesize movements from $\mathbf{q}_1$ to $\mathbf{q}_2$, and then to $\mathbf{q}_3$, we obtain the red path in Fig. 6.10, which has been computed using

$$C_{ST}(\mathbf{x}_i, \mathbf{x}_j) = r_s \arctan\left(\frac{\|\mathbf{n}_i \times \mathbf{n}_j\|}{\mathbf{n}_i \cdot \mathbf{n}_j}\right),$$

and

$$H_{ST}(\mathbf{x}_i) = c(\mathbf{x}_i, \mathbf{x}_g), \tag{6.20}$$

under the A* search strategy, where $\mathbf{n}_i = \mathbf{p}_i - \mathbf{r}_c$. Given two points $\mathbf{p}_i$ and $\mathbf{p}_j$ on the sphere, these functions provide the great-circle distance between them, so the algorithm returns motions that minimise the distance travelled by P on the surface of the sphere. Note that a simple planning approach based on interpolation in the (α_1, α_2) plane would result in a rather different motion. Indeed, the transition from

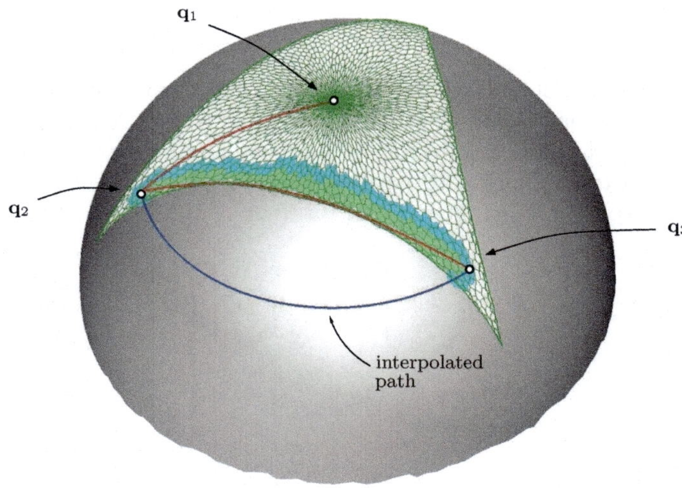

Fig. 6.10 Results of planning a path from $\mathbf{q}_1$ to $\mathbf{q}_2$, and then to $\mathbf{q}_3$, in the first experiment of Sect. 6.4.2. The figure shows the whole atlas of the wrench-feasible component attainable from $\mathbf{q}_1$ (the *green* mesh), the planned path $\mathbf{q}_1 \rightarrow \mathbf{q}_2 \rightarrow \mathbf{q}_3$ (in *red*), and the path from $\mathbf{q}_2 \rightarrow \mathbf{q}_3$ interpolated in the $(\alpha_1, \alpha 2)$ plane (in *blue*). The figure also depicts the part of the wrench-feasible C-space actually explored to plan the wrench-feasible transition from $\mathbf{q}_2$ to $\mathbf{q}_3$ (shaded in *green*, with the open charts coloured in *blue*)

$\mathbf{q}_1$ to $\mathbf{q}_2$ would coincide, but the movement from $\mathbf{q}_2$ to $\mathbf{q}_3$ would yield the blue path of the figure, which rapidly leaves $\mathcal{C}$ at the beginning.

A video in the companion web page of the book shows the results of executing both the interpolated and planned versions of the $\mathbf{q}_1 \rightarrow \mathbf{q}_2 \rightarrow \mathbf{q}_3$ movement [25]. It can be seen how, as expected, the platform moves smoothly from $\mathbf{q}_1$ to $\mathbf{q}_2$, but some cables become slack and control of the platform is lost along the interpolated path from $\mathbf{q}_2$ to $\mathbf{q}_3$. Other undesirable effects of following this path include shakyness of the platform under small perturbations, collisions with the environment, and cable tanglement at the motors. In contrast, control of the platform is maintained when following the path $\mathbf{q}_2 \rightarrow \mathbf{q}_3$ returned by the planner. Figure 6.11 summarises the video with a few snapshots.

The method can be applied to higher-dimensional problems as well. For example, if the insertion operations are to be performed with an axisymmetric tool, we can ignore the zero-torsion constraint on the platform pose by removing $\sigma = 0$ from Eq. (6.19). The result is a three-dimensional planning problem that is efficiently solved with the method, although the computation time is higher due to the increased size of the search space. Six-dimensional problems can also be solved by taking only Eq. (6.13) into account. Assuming that the sphere is no longer present, for example, a free-flying motion from $\mathbf{q}_2$ to $\mathbf{q}_3$ can be planned in a matter of seconds using the Greedy Best-First strategy. See Sect. 6.4.4.

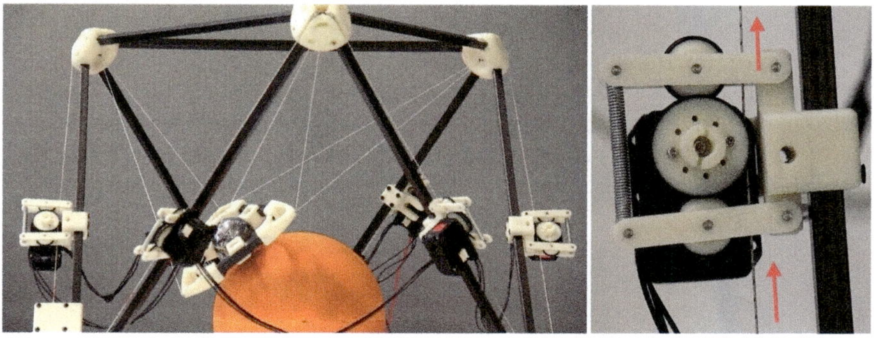

Configuration $\mathbf{q}_2$, with tight cables.

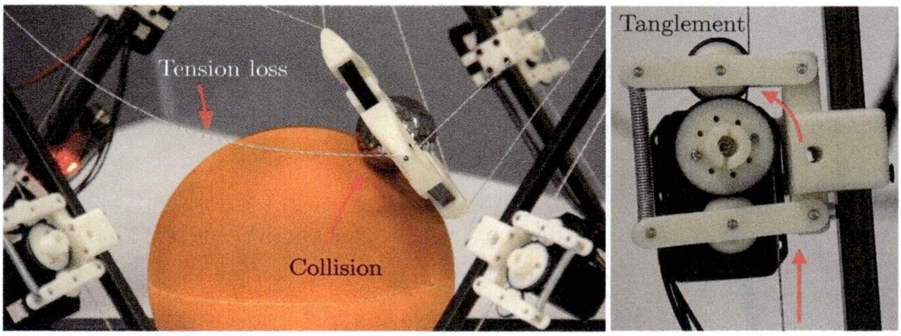

Loss of control along the interpolated path $\mathbf{q}_2 \rightarrow \mathbf{q}_3$, and cable tanglement

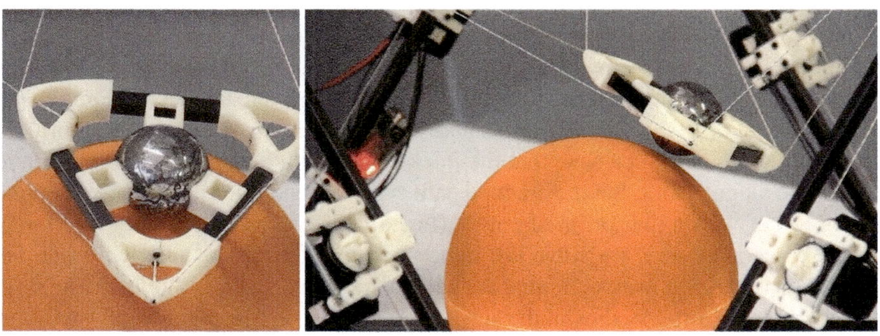

Execution of the planned path $\mathbf{q}_2 \rightarrow \mathbf{q}_3$

Fig. 6.11 Several instants of a video available in [25]. (*Top*) The robot in configuration $\mathbf{q}_2$ with all cables under tension. We see on the *right* picture that the cables correctly pass through the pulleys of the actuators. (*Middle*) Some cables lose tension and collisions appear when trying to follow the interpolated path from $\mathbf{q}_2$ to $\mathbf{q}_3$. Undesired cable tanglements also occur at the actuators. (*Bottom*) If, instead, we execute the planned path from $\mathbf{q}_2$ to $\mathbf{q}_3$, the robot undergoes a smooth motion that keeps the platform under full control

6.4.3 Planning in Rigid-Limbed Hexapods

To illustrate the performance in rigid-limbed hexapods, we adopt the geometric parameters assumed in Sect. 3.5.2, which correspond to a Gough-Stewart platform studied in [21]. In this robot, the leg lengths can vary within wide intervals, leading to a very large workspace with interesting singularity surfaces, as shown in Fig. 3.13.

Two planning queries are shown for this mechanism, using, in SI units,

$$(\underline{f_i}, \overline{f_i}) = (-300, 300),$$
$$\hat{\boldsymbol{w}}_0 = (0, 0, 1, 0, 0, 0),$$
$$\mathbf{E} = \mathbf{I}_{6\times 6},$$

where $\hat{\boldsymbol{w}}_0$ and $\mathbf{E}$ are here specified in a frame $\mathcal{F}_3$ moving parallel to $\mathcal{F}_1$, centred in the platform barycentre.

In the first query we compute a wrench-feasible path for the platform moving at the same constant orientation assumed in Sect. 3.5.2, which corresponds to the tilt-and-torsion angles

$$\boldsymbol{\tau} = (-1.449, 0.525, -1.509) \text{ rad},$$

and at a constant value of z. Using the start and goal configurations defined by the positions of P

$$\mathbf{p}_s = (0.4, 0, 0.1),$$
$$\mathbf{p}_g = (-0.3, 0, 0.1),$$

the resulting path in the (x, y) plane is computed in 578 seconds using the A* strategy. Figure 6.12 (top, left) shows this path together with the singularity curve to be avoided, the atlas corresponding to the whole wrench-feasible connected component accessible from $\mathbf{p}_s$ (shown as a mesh), and the region explored by the A* algorithm (shaded in grey). It can be seen that the interpolated path between $\mathbf{p}_s$ and $\mathbf{p}_g$ would go through singularities, but the planned path correctly encircles them while keeping the leg forces within the specified ranges (Fig. 6.12, bottom).

In the second query, we solve the same problem but only keeping constant the orientation of the platform, obtaining the path in the (x, y, z) space shown in Fig. 6.12 (top, right) in 90 min. As shown, the singularity surface, in blue, is correctly avoided by the computed path. Note that this surface is shown for illustrative purposes only, since the planner does not need to compute it explicitly to connect the start and goal configurations.

The previous two queries are hard, because the hexapod in [21] has an enormous workspace when compared to those of common rigid-limbed hexapods. To have an idea of the performance in a more usual situation, we have made additional tests using the parameters of the INRIA Left Hand [22], whose legs admit forces in the range

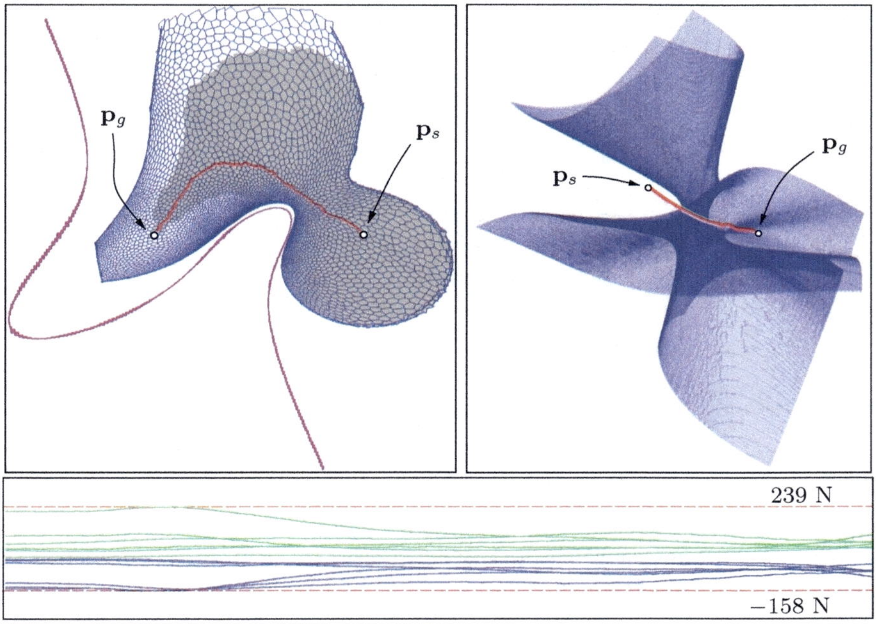

Fig. 6.12 (*Top*) Two paths computed for the mechanism in [21]. (*Bottom*) Along the *left* path, the maximum and minimum leg forces of each leg (in *green* and *blue*) stay inside the prescribed range of (−300, 300) N

$$(\underline{f_i}, \overline{f_i}) = (-300, 300)$$

and for the experiments we have set

$$\hat{\boldsymbol{w}}_0 = (0, 0, 150, 0, 0, 0),$$
$$\mathbf{E} = \mathbf{I}_{6\times 6},$$

where all has been expressed in SI units, and both $\hat{\boldsymbol{w}}_0$ and $\mathbf{E}$ are relative to a frame that translates with the platform. In this robot, the wrench-feasible C-space is close to the workspace defined by its allowable leg lengths [22], and the algorithm solves typical planning queries in a few seconds, even when permitting the variation of all pose parameters.

6.4.4 Problem Sizes and Computation Times

The problem sizes and computation times for all test cases reported in this chapter are summarised in Table 6.1. For each test case, the table shows the pose constraints assumed, the dimension of the explored manifold (n), the number of problem vari-

Table 6.1 Problem sizes and computation times (in seconds) for all test cases

Robot	Path	Pose constraints	n/m	T	
				GBF	A*
1	$\mathbf{q}_s \rightarrow \mathbf{q}_g$	x, y, z, σ fixed	2/150	29	62
	$\mathbf{q}'_s \rightarrow \mathbf{q}_g$	x, y, z, σ fixed	2/150	6	12
	$\mathbf{q}''_s \rightarrow \mathbf{q}_g$	x, y, z, σ fixed	2/150	5	9
2	$\mathbf{q}_1 \rightarrow \mathbf{q}_2$	Equation (6.19)	2/156	12	2
	$\mathbf{q}_2 \rightarrow \mathbf{q}_3$	Equation (6.19)	2/156	30	14
	$\mathbf{q}_2 \rightarrow \mathbf{q}_3$	Equation (6.19), σ free	3/157	56	166
	$\mathbf{q}_2 \rightarrow \mathbf{q}_3$	Unconstrained	6/154	40	>6000
Li et al. [21]	$\mathbf{p}_s \rightarrow \mathbf{p}_g$	ϕ, θ, σ, z fixed	2/150	70	578
	$\mathbf{p}_s \rightarrow \mathbf{p}_g$	ϕ, θ, σ fixed	3/151	90	5400
INRIA Hand [22]	Several paths	Unconstrained	6/154	<50	<100

ables involved (m), and the time in seconds (T) spent by the Greedy Best-First (GBF) and A* strategies using the cost functions explained. It becomes apparent how in terms of computation time the use of a Greedy Best-First strategy is advantageous in higher-dimensional problems, while the A* one is affordable and adviseable in lower dimensions because it normally yields lower-cost paths. Moreover, it must be said that once a partial atlas has been computed, all planning queries between configurations in this atlas can be solved in a few milliseconds.

6.5 Details About the Wrench Ellipsoid

We next provide further details on how to obtain the centre $\hat{\boldsymbol{w}}_0$ and shape matrix $\mathbf{E}$ described at the end of Sect. 6.1. To this end, we distinguish two situations, depending on whether the platform is moving in free space, or it is subject to a contact constraint. Other situations could be treated using a similar approach.

In the former case (Fig. 6.13a) the resultant wrench $\hat{\boldsymbol{w}}$ is the sum of the load weight $\hat{\boldsymbol{w}}_w$ and small perturbations introduced by inertia forces or external agents like the wind. Both $\hat{\boldsymbol{w}}_0$ and $\mathbf{E}$ are easier to describe in a frame $\mathcal{F}_3$ parallel to $\mathcal{F}_1$ in this case, located in the centre of mass G of the load. In this frame $\mathbf{E}$ is a constant matrix, and

$$\hat{\boldsymbol{w}}_0 = \hat{\boldsymbol{w}}_w = (0, 0, -w, 0, 0, 0),$$

where w is the load weight. However, since in Sect. 6.1 and Eq. (6.4) $\mathbf{J}$ was given in $\mathcal{F}_1$, both $\hat{\boldsymbol{w}}_0$ and $\mathbf{E}$ need to be converted into $\mathcal{F}_1$ for consistency. To convert $\hat{\boldsymbol{w}}_0$ we can use the transformation

$$\hat{\boldsymbol{w}}^i = \mathbf{e}_{ij} \cdot \hat{\boldsymbol{w}}^j, \tag{6.21}$$

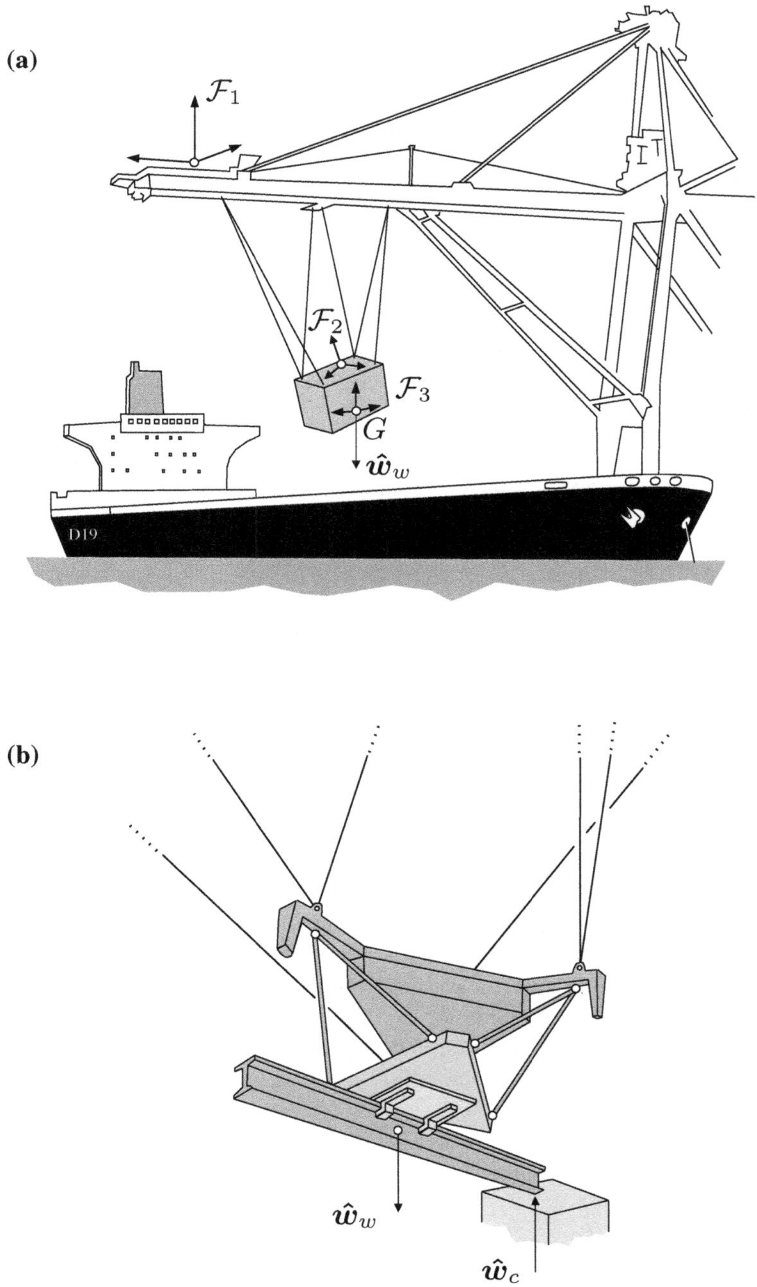

Fig. 6.13 Obtaining the wrench ellipsoid in realistic scenarios. (**a**) Cargo retrieval [17]. (**b**) Beam mounting using the NIST gripper shown in Fig. 6.2 [26]

where $\hat{\boldsymbol{w}}^i$ and $\hat{\boldsymbol{w}}^j$ indicate the coordinates of a same wrench in frames $\mathcal{F}_i$ and $\mathcal{F}_j$, and

$$\mathbf{e}_{ij} = \left[\begin{array}{c|c} \mathbf{R}_{ij} & \mathbf{0}_{3\times3} \\ \hline \mathbf{X}_{ij}\mathbf{R}_{ij} & \mathbf{R}_{ij} \end{array}\right]$$

is the change of reference matrix [6]. The block $\mathbf{R}_{ij}$ is the rotation matrix providing the orientation of $\mathcal{F}_j$ relative to $\mathcal{F}_i$, and

$$\mathbf{X_{ij}} = \begin{bmatrix} 0 & -z & y \\ z & 0 & -x \\ -y & x & 0 \end{bmatrix},$$

where x, y, and z are here the coordinates of the origin of $\mathcal{F}_j$ in frame $\mathcal{F}_i$. In Fig. 6.13a, e.g., if $\mathbf{g}$ is the position vector of G in frame $\mathcal{F}_2$ we have

$$[x, y, z]^\mathsf{T} = \mathbf{p} + \mathbf{R}\,\mathbf{g}$$

where $\mathbf{p}$ and $\mathbf{R}$ are defined as in Sect. 6.1. Similarly, to convert $\mathbf{E}$, it is easy to see that if $\mathbf{E}^i$ and $\mathbf{E}^j$ refer to the shape matrix of an ellipsoid in frames $\mathcal{F}_i$ and $\mathcal{F}_j$, then

$$\mathbf{E}^i = \mathbf{e}_{ji}^\mathsf{T} \cdot \mathbf{E}^j \cdot \mathbf{e}_{ji}, \tag{6.22}$$

where

$$\mathbf{e}_{ji} = \left[\begin{array}{c|c} \mathbf{R}_{ij}^\mathsf{T} & \mathbf{0}_{3\times3} \\ \hline \mathbf{R}_{ij}^\mathsf{T}\mathbf{X}_{ij}^\mathsf{T} & \mathbf{R}_{ij}^\mathsf{T} \end{array}\right].$$

In contact situations (Fig. 6.13b), the resultant wrench $\hat{\boldsymbol{w}}$ is the sum of the weight $\hat{\boldsymbol{w}}_w$, the contact wrench $\hat{\boldsymbol{w}}_c$, and bounded perturbations in both of such wrenches. We thus think of $\hat{\boldsymbol{w}}$ as the sum of two vectors lying inside six-dimensional ellipsoids $\mathcal{K}_w$ and $\mathcal{K}_c$ respectively, centred in $\hat{\boldsymbol{w}}_w$ and $\hat{\boldsymbol{w}}_c$, and with shape matrices $\mathbf{E}_w$ and $\mathbf{E}_c$. Therefore, $\hat{\boldsymbol{w}}$ will lie inside the Minkowski sum of $\mathcal{K}_w$ and $\mathcal{K}_c$, which, according to [27], can be tightly bounded by an ellipsoid centred in

$$\hat{\boldsymbol{w}}_0 = \hat{\boldsymbol{w}}_w + \hat{\boldsymbol{w}}_c,$$

with shape matrix

$$\mathbf{E} = \frac{1}{2}\mathbf{E}_w(\mathbf{E}_w + \mathbf{E}_c)^{-1}\mathbf{E}_c.$$

Assuming that $\hat{\boldsymbol{w}}_w$, $\mathbf{E}_w$, $\hat{\boldsymbol{w}}_c$, and $\mathbf{E}_c$ have all been expressed in frame $\mathcal{F}_1$, these are the expressions needed for $\hat{\boldsymbol{w}}_0$ and $\mathbf{E}$ in Sect. 6.1.

6.6 Extensions

In this chapter we have shown how to account for additional constraints in the problem of singularity-free path planning. For the sake of concreteness, we have described how to treat wrench-feasibility constraints in general Gough-Stewart platforms and cable-driven hexapods, but the methodology could be extended in a number of ways.

For example, it could be generalised to encompass other robot architectures because it only relies on the availability of a Jacobian relating the actuator and end-effector forces. Using the reciprocal product method outlined in Sect. 3.5.2 Zlatanov shows how to obtain a simplified velocity equation for a broad class of parallel mechanisms [28], from which such a Jacobian can be easily deduced, and more complex mechanisms could also be dealt with using the extensions presented in [29]. In particular, mechanisms with other common legs like those in Fig. 6.14 or adaptations of them could in principle be addressed, including the Hexaglide [30], mechanisms with the HEXA design [31, 32], or newer platforms like the MicARH rotary hexapod [33].

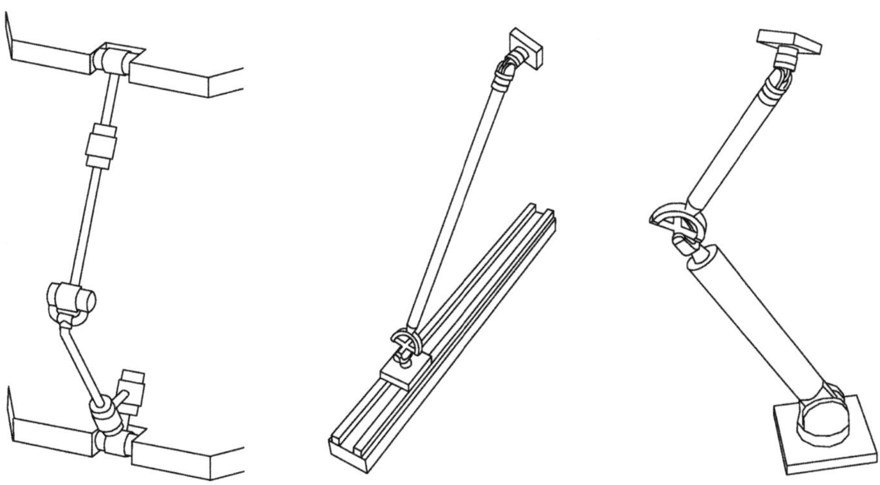

Fig. 6.14 Examples of serial kinematic chains used in parallel robots. From *left* to *right* a 6R leg, a PUS leg used in hexaglide platforms, and the RUS chain adopted for the HEXA design or the MicARH rotary hexapod. Figures adapted from [34, 35]

It would also be interesting to accommodate other kinds of constraints. For instance, one could force the positioning error of the end-effector to always stay below a given threshold, which would result in a theory reciprocal to the one that has just been presented. To impose such a constraint, we could use Eq. (3.36),

$$\mathbf{J}^{\mathsf{T}} \hat{\boldsymbol{\mathcal{T}}} = \mathbf{m}_{\upsilon},$$

which provides a relation between the twist of the end-effector, $\hat{\boldsymbol{T}}$, and the velocities of the actuators, $\mathbf{m}_{\upsilon} = \dot{\mathbf{v}}$. Note that if the pose is represented by

$$\mathbf{u} = \begin{bmatrix} \mathbf{p} \\ \boldsymbol{\tau} \end{bmatrix},$$

then $\hat{\boldsymbol{T}} = \mathbf{A}\dot{\mathbf{u}}$, where $\mathbf{A}$ is a matrix that depends on the chosen orientation parameters $\boldsymbol{\tau}$ [22, 35]. Thus, by defining $\mathbf{J}_d = \mathbf{J}^{\mathsf{T}}\mathbf{A}$ we obtain the relation

$$\mathbf{J}_d\, \dot{\mathbf{u}} = \dot{\mathbf{v}},$$

or its infinitessimals version

$$\mathbf{J}_d\, \delta\mathbf{u} = \delta\mathbf{v}$$

relating the small displacement errors in the actuated joints, $\delta\mathbf{v}$, to the corresponding errors in the pose coordinates, $\delta\mathbf{u}$. Given an uncertainty ellipsoid for $\delta\mathbf{v}$ and similarly as we did in Sect. 6.2, we could now obtain the uncertainty ellipsoid for $\delta\mathbf{u}$ and force it to remain inside a rectangular domain. We anticipate that the resulting C-space would be smooth too, which would allow us to apply the continuation method to compute paths in it.

References

1. P.A. Voglewede, I. Ebert-Uphoff, Overarching framework for measuring closeness to singularities of parallel manipulators. IEEE Trans. Robot. **21**(6), 1037–1045 (2005)
2. P. Bosscher, A.T. Riechel, I. Ebert-Uphoff, Wrench-feasible workspace generation for cable-driven robots. IEEE Trans. Robot. **22**(5), 890–902 (2006)
3. J. Hubert, *Manipulateurs Parallèles, Singularités et Analyse Statique*. PhD thesis, École Nationale Supérieure des Mines de Paris, 2010
4. I.A. Bonev, D. Zlatanov, C.M. Gosselin, Advantages of the modified Euler angles in the design and control of PKMs, in *Proceedings of the 3rd Chemnitz Parallel Kinematics Seminar/2002 Parallel Kinematic Machines International Conference (Chemnitz, Germany)* (2002), pp. 171–188
5. P. Corke, *Robotics, Vision and Control: Fundamental Algorithms in MATLAB*, vol. 73 of Springer Tracts in Advanced Robotics (Springer, 2011)
6. J.K. Davidson, K.H. Hunt, *Robots and Screw Theory: Applications of Kinematics and Statics to Robotics* (Oxford University Press, 2004)
7. K.J. Waldron, K.H. Hunt, Series-parallel dualities in actively coordinated mechanisms. Int. J. Robot. Res. **10**(5), 473–480 (1991)
8. Okuma Corporation, http://www.okuma.com/. Accessed 16 Jun 2016
9. National Institute of Standards and Technology, http://www.nist.gov. Accessed 16 Jun 2016
10. Symétrie, the Hexapod Company, http://www.symetrie.fr/en/. Accessed 16 Jun 2016
11. G.P. Assima, A. Motamed-Dashliborun, F. Larachi, Emulation of gas-liquid flow in packed beds for offshore floating applications using a swell simulation hexapod. AIChE J. **61**(7), 2354–2367 (2015)
12. Ingersoll Machine Tools, http://www.camozzimachinetools.com/. Accessed 16 Jun 2016
13. Toyoda Machinery USA, http://www.toyodausa.com/. Accessed 16 Jun 2016

14. Mikrolar, Inc., http://www.mikrolar.com/. Accessed 16 Jun 2016
15. M.E. Henderson, Multiparameter parallel search branch switching. Int. J. Bifurcat. Chaos Appl. Sci. Eng. **15**(3), 967–974 (2005)
16. S.G. Krantz, H.R. Parks, *The Implicit Function Theorem: History* ,Theory and Applications (Birkhäuser, 2002)
17. J. Albus, R.V. Bostelman, N. Dagalakis, The NIST Robocrane. J. Robotic Syst. **10**(5), 709–724 (1993)
18. S. Fang, D. Franitza, R. Verhoeven, M. Hiller, Optimum motion planning for tendon-based stewart platforms, ed. by H. Tian. *Proceedings of the 11th IFToMM World Congress in Mechanism and Machine Science (Tianjin, China)* (China Machinery Press, 2003)
19. M. Hiller, S. Fang, S. Mielczarek, R. Verhoeven, D. Franitza, Design, analysis and realization of tendon-based parallel manipulators. Mech. Mach. Theory **40**(4), 429–445 (2005)
20. S. Lahouar, E. Ottaviano, S. Zeghoul, L. Romdhane, M. Ceccarelli, Collision free path-planning for cable-driven parallel robots. Robot. Auton. Syst. **57**(11), 1083–1093 (2009)
21. H. Li, C.M. Gosselin, M.J. Richard, B.M. St-Onge, Analytic form of the six-dimensional singularity locus of the general Gough-Stewart platform. ASME J. Mech. Des. **128**(1), 279–287 (2006)
22. J.-P. Merlet, *Parallel Robots* (Springer, 2006)
23. The CUIK Project Home Page, http://www.iri.upc.edu/cuik. Accessed 16 Jun 2016
24. D.Q. Huynh, Metrics for 3D rotations: comparison and analysis. J. Math. Imaging Vis. **35**(2), 155–164 (2009)
25. Companion web page of this book, http://www.iri.upc.edu/srm. Accessed 16 Jun 2016
26. A.M. Lytle, K.S. Saidi, NIST research in autonomous construction. Auton. Robots **22**(3), 211–221 (2007)
27. L. Ros, A. Sabater, F. Thomas, An ellipsoidal calculus based on propagation and fusion. IEEE Trans. Syst. Man, Cybern. Part B: Cybern. **32**(4), 430–442 (2002)
28. D. Zlatanov, *Generalized Singularity Analysis of Mechanisms*. PhD thesis, University of Toronto, 1998
29. M. Zoppi, D. Zlatanov, R. Molfino, On the velocity analysis of interconnected chain mechanisms. Mech. Mach. Theory **41**(11), 1346–1358 (2006)
30. M. Honegger, A. Codourey, E. Burdet, Adaptive control of the hexaglide, a 6 DOF parallel manipulator, in *Proceedings of the IEEE International Conference on Robotics and Automation, ICRA (Albuquerque, USA)*, (1997), pp. 543–548
31. H. Bruyninckx, The 321-HEXA: a fully-parallel manipulator with closed-form position and velocity kinematics, in *Proceedings of the IEEE International Conference on Robotics and Automation, ICRA (Albuquerque, USA)* (1997), pp. 2657–2662
32. F. Pierrot, P. Dauchez, A. Fournier, HEXA: a fast six-DOF fully-parallel robot. in *Fifth International Conference on Robots in Unstructured Environments*, vol. 2, (1991), pp. 1158–1163
33. J. Coulombe, I.A. Bonev, A new rotary hexapod for micropositioning, in *Proceedings of the IEEE International Conference on Robotics and Automation, ICRA (Karlsruhe, Germany)* (2013), pp. 877–880
34. J. Angeles, *Fundamentals of Robotic Mechanical Systems: Theory, Methods, and Algorithms* (Springer, 2006)
35. H. Bruyninckx, J. de Shutter, Introduction to intelligent robotics, Technical Reports, Katholieke Universiteit de Leuven, 2001

Chapter 7
Conclusions

This book has presented new methods for singularity set computation and singularity-free path planning. Unlike in previous works, the two problems have been addressed assuming a general kinematic architecture and geometry of the robotic systems under consideration. By design, the methods are applicable to any nonredundant mechanism with lower-pair joints, the only limitation coming from the computational power available. Special emphasis has been put on illustrating the methods in mechanisms with complex architectures because they are those typically arising in today's robotics, where the development of new machines with increasingly complex motions is a growing trend. To conclude the book, we summarise the presented results and highlight points for future attention.

7.1 Summary of Results

The book has contributed with the following methods:

A general method for singularity set computation (Chaps. 2 and 3)

The method can isolate the whole singularity set independently of its dimension, even in the presence of bifurcations, sharpnesses, or dimension changes. Likewise, the method can compute any of the singularity subsets typically defined, including the forward, inverse, or C-space singularity sets, or any of the low-level subsets defined by Zlatanov [1]. An effort has also been made to provide guidelines on how to represent the various singularity sets once obtained, aiming to produce suitable diagrams for the robot designer. On this regard, it has been shown how the set projections to the input or output spaces provide global information about the motion capabilities of the mechanism, including the reachable areas, the locations where rigidity or dexterity losses arise, and the regions where mechanism motions can safely be planned.

A general method for workspace determination (Chap. 4)

The proposed method provides a detailed map of the workspace regions swept by any set of coordinates, indicating all motion impediments, or barriers, that may be encountered in their interior. This is a substantially richer output than that produced by previous methods for particular mechanisms. In comparison to general methods

O. Bohigas et al., *Singularities of Robot Mechanisms*,
Mechanisms and Machine Science 41, DOI 10.1007/978-3-319-32922-2_7

based on continuation, the method is advantageous in that it is *uninformed*, since it does not need to be fed with precomputed assembly configurations or suitable slicing directions, and *complete*, as it returns the full workspace even in the presence of several connected components, hidden regions, or degenerate barriers. The method is also able to compute workspaces of any dimension, but those of dimension larger than three will be hard to compute, as in any other method. Since such workspaces are not directly visualisable either, the common practice is to obtain three-dimensional subsets meaningful to the robot designer, such as the reachable, constant-orientation, and constant-position workspaces, which can all be computed by the proposed method. Furthermore, while many previous methods are limited to obtain cross-sectional curves of the workspace boundary, the one we present will isolate the full boundary surfaces directly.

A general singularity-free path planning method (Chap. 5)

Due to the complexity of the involved C-spaces and their singularity loci, previous methods have only considered C-spaces admitting explicit parameterisations. In contrast, the presented approach makes no recourse to such parameterisations, which makes it applicable to nonredundant lower-pair mechanisms of arbitrary architecture. The planning method relies on a system of equations that implicitly characterises the singularity-free C-space of the mechanism, which avoids the need of representing the singularity locus explicitly as an obstacle. The solution manifold of this system can be freely navigated without fear of crossing any singularity of the mechanism. Higher-dimensional continuation techniques are then used to progressively construct an atlas of the connected component containing the start configuration, until the goal configuration is reached, or path nonexistence is proved at the resolution of the search. The method can also be used to detect nonsingular transitions between forward kinematic solutions [2], and it can generate an exhaustive atlas of the singularity-free component that is reachable from one configuration. The latter ability is useful to rapidly resolve subsequent planning queries, or to visualise the singularity-free workspace relative to any set of mechanism coordinates.

A method to plan paths that keep a physically-meaningful clearance with respect to the singularity locus (Chap. 6)

The approach is based on constraining the actuator forces to remain within the desired bounds, for a given end-effector wrench subject to six-dimensional uncertainty. In doing so, we guarantee the avoidance of forward singularities, and at the same time obtain a physically-intuitive way to tune the clearance with respect to the singularity locus. To increase or decrease the clearance we only have to narrow or widen the force bounds of the actuators, respectively. By way of example, the approach has been explained for rigid-limbed or cable-driven hexapods, but hints have also been given about how to extend the planner to cope with other mechanisms or clearance constraints.

7.2 Future Research Directions

As in any research, our study opens up more questions than it answers. We next briefly discuss a few points that would deserve further attention.

Exploitation of Symmetries

One of the most interesting extensions of this work would we the development of new tools and algorithms to explicitly deal with mechanism symmetries. Note that robot designers often tend to define symmetric mechanism designs, and a quick look to the test cases of the book reveals that such designs also yield symmetric singularity loci. If for example we project the 3-D singularity sets shown in Figs. 5.11 and 5.12 vertically to the (x_7, y_7) plane, we obtain the results in Fig. 7.1, which illustrate this point. If a mechanism exhibits a given group of symmetry operations, what symmetries are to be expected in its configuration space and in the associated singularity loci? Can we predict them a priori? Can we restrict the computations of a given singularity set to a certain "fundamental region" that determines, by symmetry, the shape of the remaining regions of the set? If that would be possible, it would probably reduce substantially the computational effort required to compute the various singularity sets. The work by Sljoka, Schulze, and Whiteley is quite relevant in pursuing such an extension [3–7].

Redundant Mechanisms

A natural continuation of this research would be to extend the developments to also account for redundant mechanisms. Their treatment was left out of the scope of the present work, but Zlatanov already provides definitions and classification methods for the singularities of such mechanisms [1], which would constitute an ideal basis from which to construct such an extension.

Collision Constraints

When computing singularities, collision constraints between the various bodies of the mechanism or with the environment have been neglected, and they have only been treated slightly by the planners proposed in Chaps. 5 and 6, as penalty terms in the cost functions considered. A subject of further investigation might be the formulation of such constraints with equations of a proper form, in order to integrate them into Eq. (2.1). In this way, it would be possible to compute the singularity loci lying in the C-space regions compatible with such constraints, and to directly plan collision-free movements in the resulting manifolds.

Detailed Singularity Analysis of Complex Spatial Mechanisms

The interpretation and visualisation of the singularity sets requires a profound and detailed analysis that can now be confronted using the methods explained in the book. It would be interesting to apply the presented algorithms to the analysis of additional or more challenging mechanisms.

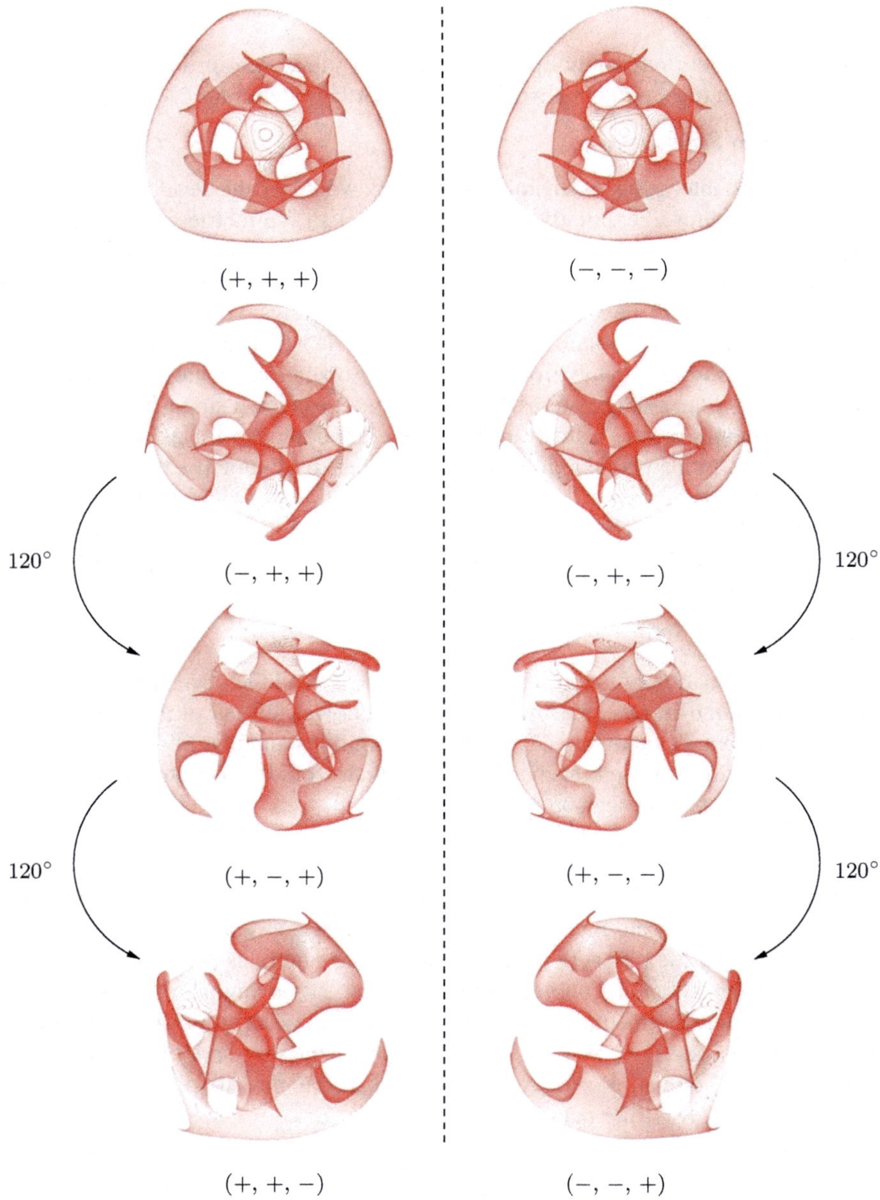

Fig. 7.1 The surfaces in Figs. 5.11 and 5.12 vertically projected to the (x_7, y_7) plane. The *left* and *right* working modes are pairwise mirror symmetric. The last three modes in each column exhibit a 120° rotational symmetry

Stratification of the Singularity Set

As noted in [1], one would like to obtain a stratification of the singularity set into nonintersecting submanifolds of a same class. This could be relevant in the early design stages of a mechanism, to identify the exact physical phenomena that occur at singular configurations. Preliminary work along this direction shows that our developments can be used to design a hierarchy of tests for the whole singularity set, which yields sets of classes to which an identified singularity cannot belong [8]. The improvement of this approach and its application to spatial mechanisms demands additional work.

Dexterous Workspaces

A class of workspaces not treated explicitly in this book are dexterous workspaces, which are defined as the set of end-effector positions that can be reached for *any* orientation in a given range. While computing such workspaces seems plausible under the proposed approach, several modifications need to be introduced in order to ultimately achieve so. Mathematical conditions allowing the identification of the boundary of such workspaces have already been investigated [9–14], which could be used as a starting point for the development of this extension.

Variants of the Path Planners

The resolution completeness of the path planners proposed comes at the expense of a computational cost that scales exponentially with the dimension of the explored manifold. To deal with higher-dimensional problems we could adapt the randomised approach in [15], which trades off resolution completeness for efficiency and probabilistic completeness, or the approach in [16], which additionally guarantees asymptotic optimality. The evaluation of these variants in the context of singularity-free or wrench-feasible path planning certainly deserves further attention, but from our experience we foresee the solvability of problems in dimensions six and above in reasonably-short times.

Clearance Relative to Singularities

A point left open in Chap. 5 is how to establish the value of the threshold b_{max} governing the clearance of the path relative to the singularity locus. As mentioned, this threshold has to be determined case-wise on each particular context of application, but specific versions of the planner for particular mechanisms may lead to systematic ways to set a meaningful clearance. For example, Chap. 6 has shown how to do so in hexapod mechanisms using the idea that the leg forces should remain within the allowed limits. It should be possible to develop a dual strategy that fixes the clearance in terms of a maximum position/orientation error permitted to the platform, which would be nice to apply to mechanisms designed for fine-positioning applications [17–19].

A General Wrench-Feasible Path Planner

The wrench-feasible path planner of Chap. 6 should be extended to also deal with other mechanism architectures. Note that the method depends on the availability of a force Jacobian relating the platform wrench with the joint forces or torques. In parallel mechanisms, for example, this Jacobian can be obtained using the reciprocal screw method developed in [1, Chap. 4], but larger mechanism classes should be addressable using the results in [20].

Dynamical Effects

Some mechanisms, like those used for fast pick-and-place operations, are designed to be operated at very high speeds. Thus, accounting for dynamical effects in the proposed path planners would be of great interest [21]. In particular, further research should be conducted to allow the computation of the trajectory that minimises the travel time between the query configurations, as done in [22] for the 2-$\underline{R}$RR robot DexTAR for example.

References

1. D. Zlatanov, Generalized singularity analysis of mechanisms. PhD thesis, University of Toronto, 1998
2. O. Bohigas, M.E. Henderson, L. Ros, J.M. Porta, A singularity-free path planner for closed-chain manipulators, in *Proceedings of the IEEE International Conference on Robotics and Automation, ICRA (St. Paul, USA)* (2012), pp. 2128–2134
3. B. Schulze, Symmetry as a sufficient condition for a finite flex. SIAM J. Discrete Math. **24**(4), 1291–1312 (2010)
4. B. Schulze, W. Whiteley, The orbit rigidity matrix of a symmetric framework. Discrete Comput. Geom. **46**(3), 561–598 (2011)
5. A. Sljoka, Algorithms in rigidity theory with applications to protein flexibility and mechanical linkages. PhD Thesis, York University, 2012
6. B. Schulze, A. Sljoka, W. Whiteley, How does symmetry impact the rigidity of proteins? Philos. Trans. Ser. A, Math. Phys. Eng. Sci. **372**(2008), 20120041 (2014)
7. J.M. Porta, L. Ros, B. Schulze, A. Sljoka, W. Whiteley, On the symmetric molecular conjectures, in *Computational Kinematics*, ed. by F. Thomas, A. Pérez Gracia (Springer 2014), pp. 175–184
8. O. Bohigas, D. Zlatanov, M. Manubens, L. Ros, On the numerical classification of the singularities of robot manipulators, in *Proceedings of the ASME International Design Engineering Technical Conferences and Computers and Information in Engineering Conference, IDETC/CIE (Chicago, USA)* (2012), pp. 1287–1296
9. E.J. Haug, J.-Y. Wang, J.K. Wu, Dextrous workspaces of manipulators, part I: Analytical criteria. J. Struct. Mech. **20**(3), 321–361 (1992)
10. J.-Y. Wang, J.K. Wu, Dextrous workspaces of manipulators, part II: Computational methods. J. Struct. Mech. **21**(4), 471–506 (1993)
11. C.C. Qiu, C.-M. Luh, E.J. Haug, Dextrous workspaces of manipulators, part III: Calculation of continuation curves at bifurcation points. J. Struct. Mech. **23**(1), 115–130 (1995)
12. L.J. Du Plessis, J.A. Snyman, A numerical method for the determination of dextrous workspaces of Gough-Stewart platforms. I. J. Numer. Methods Eng. **52**(4), 345–369 (2001)
13. A.M. Hay, J.A. Snyman, A multi-level optimization methodology for determining the dextrous workspaces of planar parallel manipulators. Struct. Multi. Optim. **30**(6), 422–427 (2005)
14. F.-C. Yang, E.J. Haug, Numerical analysis of the kinematic dexterity of mechanisms. ASME J. Mech. Des. **116**(1), 119–126 (1994)
15. L. Jaillet, J.M. Porta, Path planning under kinematic constraints by rapidly exploring manifolds. IEEE Trans. Robot. **29**(1), 105–117 (2013)

16. L. Jaillet, J.M. Porta, Asymptotically-optimal path planning on manifolds, in *Robotics: Science and Systems* (2012)
17. L. Campos, F. Bourbonnais, I.A. Bonev, P. Bigras, Development of a five-bar parallel robot with large workspace, in *Proceedings of the ASME International Design Engineering Technical Conferences and Computers and Information in Engineering Conference, IDETC/CIE (Montreal, Canada)* (2010)
18. F. Bourbonnais, Utilisation optimale de l'espace de travail des robots parallèles en affrontant certains types de singularités," Master's thesis, École de Technologie Supérieure, Université du Québec (2012)
19. J. Coulombe, I.A. Bonev, A new rotary hexapod for micropositioning, in *Proceedings of the IEEE International Conference on Robotics and Automation, ICRA (Karlsruhe, Germany)* (2013), pp. 877–880
20. M. Zoppi, D. Zlatanov, R. Molfino, On the velocity analysis of interconnected chain mechanisms. Mech. Mach. Theory **41**(11), 1346–1358 (2006)
21. A. Müller, On the terminology and geometric aspects of redundant parallel manipulators. Robotica **31**(1), 137–147 (2013)
22. F. Bourbonnais, P. Bigras, I. Bonev, Minimum-time trajectory planning and control of a pick-and-place five-bar parallel robot. IEEE/ASME Trans. Mechatron. **20**(2), 740–749 (2015)

Author Index

O. Bohigas et al., *Singularities of Robot Mechanisms*,
Mechanisms and Machine Science 41, DOI 10.1007/978-3-319-32922-2

Z

Subject Index

O. Bohigas et al., *Singularities of Robot Mechanisms*,
Mechanisms and Machine Science 41, DOI 10.1007/978-3-319-32922-2

Printed by Printforce, the Netherlands